NOTATION

Letters

A	=	cross-sectional area
c'	=	effective cohesion
C_c	=	compression index
C_r	=	recompression index
c_v	=	coefficient of vertical consolidation
D	=	particle size
e	=	void ratio
F	=	shear force
g	=	gravitational acceleration constant
G_s	=	specific gravity of soil solids
h	=	head
i	=	hydraulic gradient
k	=	hydraulic conductivity
LL	=	liquid limit
M	=	total mass of soil
M_s	=	mass of solids in soil
M_w	=	mass of water in soil
n	=	porosity
N	=	normal force
P	=	percent of soil solids finer than D
PI	=	plasticity index
PL	=	plastic limit
q	=	flow rate
Q	=	flow volume
q_u	=	unconfined compressive strength
S	=	degree of saturation
s_u	=	undrained shear strength
t	=	time
u	=	pore pressure
V	=	total volume of soil
V_s	=	volume of solids in soil
V_v	=	volume of voids in soil
V_w	=	volume of water in soil
v_D	=	Darcian velocity
v_s	=	seepage velocity
W	=	total weight of soil
w	=	moisture content
w_{opt}	=	optimum moisture content
W_s	=	weight of solids in soil
W_w	=	weight of water in soil

Symbols

$\Delta\sigma$	=	deviator stress
ε	=	axial strain
ϕ'	=	effective friction angle
γ	=	total unit weight
γ_d	=	dry unit weight
γ_{dmax}	=	dry unit weight corresponding to w_{opt}
γ_w	=	unit weight of water
σ	=	total stress
σ'	=	effective stress
σ_1	=	major principal stress
σ_3	=	minor principal stress
σ'_{max}	=	max. previous consolidation pressure
τ	=	shear stress
τ_f	=	shear strength

SOIL MECHANICS LAB MANUAL

2nd Edition

Michael E. Kalinski, Ph.D., P.E.
University of Kentucky

JOHN WILEY & SONS, INC.

Copyright © 2011 by John Wiley & Sons, Inc.

Founded in 1807, John Wiley & Sons, Inc. has been a valued source of knowledge and understanding for more than 200 years, helping people around the world meet their needs and fulfill their aspirations. Our company is built on a foundation of principles that include responsibility to the communities we serve and where we live and work. In 2008, we launched a Corporate Citizenship Initiative, a global effort to address the environmental, social, economic, and ethical challenges we face in our business. Among the issues we are addressing are carbon impact, paper specifications and procurement, ethical conduct within our business and among our vendors, and community and charitable support.

For more information, please visit our website: www.wiley.com/go/citizenship.

ISBN-13   978-0-470-55683-2

Printed in the United States of America

10 9 8 7 6 5 4 3 2 1

Printed and bound by Hamilton Printing Company

TABLE OF CONTENTS

PREFACE

This manual is written for the laboratory component of a typical one-semester undergraduate soil mechanics course as part of a typical civil engineering undergraduate curriculum. The manual is written as a stand-alone document, but supporting media have also been prepared to enhance the learning process. These resources are available online at www.wiley.com/college/kalinski, and include the following:

- **Video Demonstrations.** Brief (10-20 minutes) video demonstrations have been produced for each laboratory test. Each video describes the basic purpose of the test, lists the required materials, demonstrates the step-by-step procedure, and details methods for reducing the data. Viewing these videos prior to the lab will help prepare the students for the lab exercise, and ultimately enhance the students' learning experiences.

- **Laboratory Data Sheets.** Generic laboratory data sheets have been prepared for each exercise, and are included at the end of each chapter. These data sheets are intended for use by students, researchers, or practicing engineers. These forms can also be downloaded off of the website listed above.

ALSO AVAILABLE FROM WILEY:

Soil Mechanics and Foundations, 2nd Edition, by Muniram Budhu
ISBN: 0-471-43117-6
web: www.wiley.com/college/budhu

If you would like to learn more about the concepts and fundamental principles behind soil mechanics, Muniram Budhu of the University of Arizona has written an introductory text for soil mechanics and foundations. This book is written for soil mechanics courses typically offered as part of undergraduate civil engineering curricula. The book includes numerous solved example problems and homework exercises. An accompanying CD-ROM integrates interactive animations, interactive problem solving, interactive step-by-step examples, a virtual soils laboratory, and e-quizzes to engage student learning and retention.

Michael E. Kalinski
University of Kentucky

ACKNOWLEDGMENTS

This soil mechanics laboratory manual was inspired by the undergraduate students that I have taught over the past 15 years at the University of Texas at Austin and the University of Kentucky. They taught me what is important and what is effective with respect to laboratory instruction, and those lessons have helped to shape this manual. To them, I express my utmost gratitude.

I would also like to extend my gratitude to all of my friends and colleagues who have helped to make this manual possible. Dave Daniel, Roy Olson, Ken Stokoe, Priscilla Nelson, and Steven Wright at the University of Texas at Austin inspired me as a graduate student to learn about soil mechanics and become an instructor. Bobby Hardin, Issam Harik, and Jerry Rose provided ample guidance and encouragement to me as a young professor here at UK. Erwin Supranata provided valuable input and suggestions as a graduate student at UK working in the soils lab. Bettie Jones, Jim Norvell, Ruth White, Shelia Williams, and Gene Yates have provided administrative assistance and support in the lab, without which this manual would not have been possible. Darchelle Leggett, Mary Moran, Wendy Perez, and Jenny Welter provided guidance and encouragement to help me through the publication process with Wiley. Seven of my colleagues who reviewed the manual, including Joe Caliendo, Jeffrey Evans, and Robert Johnson, provided constructive criticism and suggestions that greatly enhanced the quality and usefulness of this manual. Terry Edin, Kelan Griffin, and Stuart Reedy provided valuable assistance and resources at UK during the production of the video demonstrations that accompany this manual.

Finally, I would like to thank my family: Pamela, Jackson, and Lucas, for bringing me happiness every day. This manual is dedicated to them.

Michael E. Kalinski
University of Kentucky

1. INTRODUCTION

1.1. THE IMPORTANCE OF LABORATORY SOIL MECHANICS TESTING

Soil can exist as a naturally occurring material in its undisturbed state, or as a compacted material. Geotechnical engineering involves the understanding and prediction of the behavior of soil. Like other construction materials, soil possesses mechanical properties related to strength, compressibility, and permeability. It is important to quantify these properties to predict how soil will behave under field loading for the safe design of soil structures (e.g. embankments, dams, waste containment liners, highway base courses, etc.), as well as other structures that will overly the soil. Quantification of the mechanical properties of soil is performed in the laboratory using standardized laboratory tests.

1.2. OVERVIEW OF MANUAL CONTENTS

The main objectives of a laboratory course in soil mechanics are to introduce soil mechanics laboratory techniques to civil engineering undergraduate students, and to familiarize the students with common geotechnical test methods, test standards, and terminology. The procedures for all of the tests described in this manual are written in accordance with applicable American Society for Testing and Materials (ASTM) standards. It is important to be familiar with these standards to understand, interpret, and properly apply laboratory results obtained using a standardized method. Each test described in this manual has an associated ASTM standard number as summarized in Table 1.1.

Each chapter in the manual describes one test, but the instructor may choose to combine more than one test during a given laboratory session. For example, the moisture content and specific gravity laboratory exercises are relatively short, so it would be reasonable to combine these exercises into one three-hour laboratory period. Each chapter is structured in the same manner, and includes the following sections:

- Section 1 – Applicable ASTM Standards;
- Section 2 – Purpose of Measurement;
- Section 3 – Definitions and Theory;
- Section 4 – Equipment and Materials;
- Section 5 – Procedure;
- Section 6 – Expected Results (for quantitative measurements);
- Section 7 – Likely Sources of Error;
- Section 8 – Additional Considerations; and
- Section 9 – Suggested Exercises.

Laboratory data sheets are included at the end of each chapter. Data sheets are written to be used for practical purposes as well as educational purposes, with places to insert information regarding project, boring number, and soil Recovery Depth/Method.

1

Additional data sheets can be found on the companion website that accompanies this manual (www.wiley.com/college/kalinski). When accessing the website, you will need your registration code, which can be found on the card inside the envelope just inside the front cover of the manual.

Table 1.1—List of laboratory exercises and applicable ASTM standards

Laboratory Exercise	Chapter	Applicable ASTM Standard(s)
Moisture Content of Soil	2	D2216
Specific Gravity of Soil Solids	3	D854
Liquid Limit and Plastic Limit of Soil	4	D4318
Analysis of Grain Size Distribution	5	D422, D1140
Laboratory Classification of Soil	6	D2488
Field Classification of Soil	7	D2487
Laboratory Soil Compaction	8	D698, D1557
Field Measurement of Dry Unit Weight	9	D1556, D2167
Hydraulic Conductivity of Granular Soil Using a Fixed Wall Permeameter	10	D2434
One-Dimensional Consolidation Test of Cohesive Soil	11	D2435
Direct Shear Strength Test of Granular Soil	12	D3080
Unconfined Compressive Strength Test	13	D2166
Unconsolidated-Undrained Triaxial Shear Strength Test of Cohesive Soil	14	D3018

1.3. REVIEW OF WEIGHT-VOLUME RELATIONSHIPS IN SOILS

Soil is a porous medium consisting of soil solids (mineral grains) and voids. Some of the voids are filled with air, and some are filled with water. The different components of soil (soil solids, water-filled voids, and air-filled voids) each possess weight and volume as defined in Fig. 1.1.

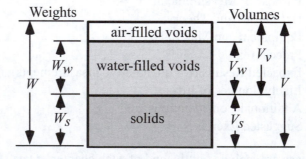

Fig. 1.1—Definitions of parameters used for weight-volume calculations in soil.

Throughout this manual, you will be required to perform weight-volume calculations of soil. Discussion of weight-volume relationships (a.k.a. phase relationships) is standard material for undergraduate soil mechanics lecture courses, but is also included in this manual for your information. This review does not present an exhaustive list of equations for you to remember. It simply includes a "toolbox" of basic definitions and relationships that you can use to perform most weight-volume relationship calculations. In soil mechanics, we define several terms based on the parameters shown in Fig. 1.1. These terms form the basis for weight-volume calculations, and are defined in Table 1.2.

Table 1.2—Basic terms used in weight-volume relationships in soil.

Term	Equation	Typical Range in Soil
Total Unit Weight	$\gamma = \dfrac{W}{V}$	90-140 lbs/ft^3 (pcf)
Dry Unit Weight	$\gamma_d = \dfrac{W_s}{V}$	80-130 pcf
Moisture Content	$w = \dfrac{W_w}{W_s} \; x \; 100\%$	10-50%
Unit Weight of Water	$\gamma_w = \dfrac{W_w}{V_w}$	62.4 pcf
Specific Gravity of Soil Solids	$G_s = \dfrac{W_s}{\gamma_w V_s}$	2.65-2.80
Void Ratio	$e = \dfrac{V_v}{V_s}$	0.3-1.5
Porosity	$n = \dfrac{V_v}{V} \; x \; 100\%$	25-60%
Degree of Saturation	$S = \dfrac{V_w}{V_v} \; x \; 100\%$	10-100%

1.3.   PREPARATION OF PROFESSIONAL-QUALITY GRAPHS

Many students have difficulty creating professional-quality graphs of experimental data simply because they have not received any formal guidance and instruction. With the widespread use of commercial graphics and spreadsheet software to create graphs, many students just assume the computer will automatically create an acceptable graph with the given data. However, this is usually not the case. One goal of this laboratory is to teach students how to present experimental data in a professional manner. An acceptable graph must satisfy all of the following criteria:

- Title that describes the test performed and the data presented;
- Date and name of creator;

- Major axes at a sensible interval;
- Use of appropriate scale (either logarithmic or linear);
- Axes labeled and units given; and
- Data that fill up most of the graph space.

Examples of acceptable and unacceptable graphs are shown in Figs. 1.2 and 1.3.

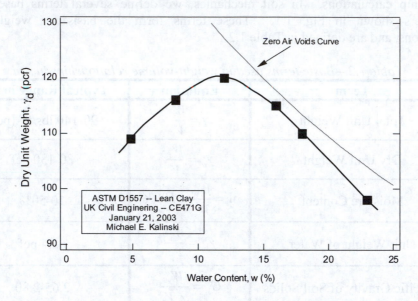

Fig. 1.2—Example of an acceptable graph.

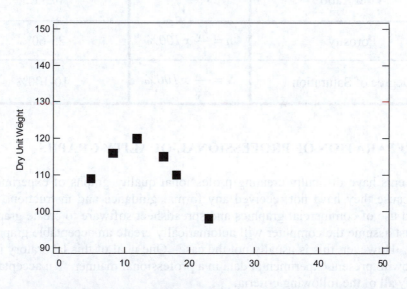

Fig. 1.3—Example of an unacceptable graph (axis label missing, units missing, graph title missing, and data do not fill the graph space).

When used properly, commercial software is a very valuable tool for graphically presenting data. When using commercial software, be careful when applying any automatic curve-fitting utility. Students often use this utility to obtain nonsensical results, which they blindly submit as part of their laboratory report without considering the validity of the curve fit. If an automatic curve-fitting utility is used, you should always check the curve fit against the expected trend.

1.4. VIDEO DEMONSTRATIONS

Brief video demonstrations of each lab can be found on the companion website that accompanies this manual (www.wiley.com/college/kalinski). When accessing the website, you will need your registration code, which can be found on the card inside the envelope just inside the front cover of the manual. Each demonstration includes a brief background of the test, required equipment, and step-by-step procedure for the measurement and reduction of experimental data. These demonstrations are not intended to replace the demonstrations and guidance provided by your laboratory instructor, but are merely intended to serve as a supplement to your educational experience. Nevertheless, it is recommended that you take the time to view each demonstration prior to the laboratory.

2. MEASUREMENT OF MOISTURE CONTENT

2.1. APPLICABLE ASTM STANDARD

- ASTM D2216: Standard Test Method for Laboratory Determination of Water (Moisture) Content of Soil and Rock by Mass

2.2. PURPOSE OF MEASUREMENT

Moisture content measurement is primarily used for performing weight-volume calculations in soils. Moisture content is also a measure of the shrink-swell and strength characteristics of cohesive soils as demonstrated in liquid limit and plastic limit testing.

2.3. DEFINITIONS AND THEORY

The mass of a given volume of moist soil is the sum of the mass of soil solids, M_s, and the mass of water in the soil, M_w. Moisture content, w, is defined as:

$$w = \frac{M_w}{M_s} \, x \, 100\% .$$

(2.1)

Moisture content is typically expressed as a percentage using two significant figures (e.g. 12%, 9.2%, etc.). Moisture content can range from a few percent for "dry" sands to over 100% for highly plastic clays. Even soils that appear to be "dry" possess some moisture.

2.4. EQUIPMENT AND MATERIALS

The following equipment and materials are required for moisture content measurements:

- Disturbed sample of moist soil;
- scale capable of measuring to the nearest 0.01 g;
- soil drying oven set at $110^\circ \pm 5^\circ$ C;
- 3 oven-safe containers; and
- permanent marker for labeling containers.

2.5. PROCEDURE[1]

The moisture content calculation is based on three measurements:

---

[1] Don't forget to visit www.wiley.com/college/kalinski to view the lab demo!

1) Mass of container, M_c;
2) mass of moist soil plus container before drying, M_1; and
3) mass of dry soil plus container after drying, M_2.

Moist soil is placed in an oven-safe container and dried for 12-16 hours in a soil drying oven. It is helpful to use an oven mitt or tongs to insert and remove the containers from the oven. The soil-filled container is weighed before and after drying to obtain M_1 and M_2, respectively, and w is calculated as:

$$w = \frac{M_w}{M_s} \, x \, 100\% = \frac{M_1 - M_2}{M_2 - M_c} \, x \, 100\% .$$
(2.2)

2.6. EXPECTED RESULTS

In coarse-grained soils such as sands and gravels, w may range from a few percent in drier soils to over 20% in saturated soils. In fine-grained soils such as silts and clays, the possible range in w is much higher due to the ability of clay minerals to adsorb water molecules. Moisture content in fine-grained soils may be as low as a few percent, to over 100% in higher-plasticity clays.

2.7. LIKELY SOURCES OF ERROR

For moisture content measurement, likely sources of error may include inadequate drying, or excessive drying beyond the recommended 12-16 hour drying period. According to ASTM D2216, soil should be dried at 110°C for 12-16 hours. However, for soils containing a significant amount of organic material or hydrous minerals such as gypsum, some of the water is bound by the soil solids, so excessive drying will effectively drive some of the soil solids away and produce erroneous results. In these cases, the oven temperature should be reduced to 60°C.

2.8. ADDITIONAL CONSIDERATIONS

With respect to moisture content measurements and specimen size, the recommended amount of soil required to obtain an accurate measurement increases with increasing maximum particle size, with a minimum of 20 g, as shown in Fig. 2.1.

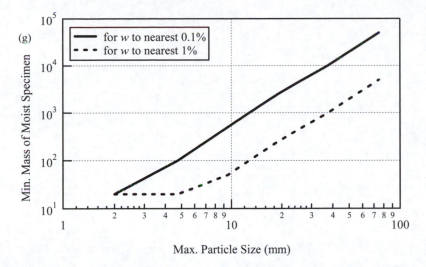

Fig. 2.1—Recommended minimum sample mass for moisture content testing based on maximum particle size.

2.9.   SUGGESTED EXERCISES

1)   Perform moisture content measurement of three specimens of the soil supplied by your instructor, and present your results using the Measurement of Moisture Content Laboratory Data Sheet at the end of this chapter (additional data sheets can be found on the CD-ROM that accompanies this manual).

2)   What temperature should be used to dry most soil specimens for moisture content measurement? What exceptions exist, and what temperatures should be used for those exceptions?

3)   How long should most specimens be dried to obtain an accurate moisture content measurement?

MEASUREMENT OF MOISTURE CONTENT (ASTM D2216)
LABORATORY DATA SHEET

I. GENERAL INFORMATION

Tested by:	Date tested:
Lab partners/organization:	
Client:	Project:
Boring no.:	Recovery depth:
Recovery date:	Recovery method:
Soil description:	

II. TEST DETAILS

Oven temperature:	Drying time:
Scale type/precision/serial no.:	
Notes, observations, and deviations from ASTM D2216 test standard:	

III. MEASUREMENTS AND CALCULATIONS

Container ID:			
Mass of container (M_c):			
Mass of moist soil + container (M_1):			
Mass of dry soil + container (M_2):			
Mass of moisture (M_w):			
Mass of dry soil (M_s):			
Moisture content (w):			
Average moisture content:			

IV. EQUATION AND CALCULATION SPACE

$$w = \frac{M_1 - M_2}{M_2 - M_c} \times 100\%$$

3. MEASUREMENT OF SPECIFIC GRAVITY OF SOIL SOLIDS

3.1. APPLICABLE ASTM STANDARD

- ASTM D854: Standard Test Method for Specific Gravity of Soils

3.2. PURPOSE OF MEASUREMENT

Specific gravity of soil solids is used for performing weight-volume calculations in soils.

3.3. DEFINITIONS AND THEORY

Specific gravity of soil solids, G_s, is the mass density of the mineral solids in soil normalized relative to the mass density of water. Alternatively, it can be viewed as the mass of a given volume of soil solids normalized relative to the mass of an equivalent volume of water. Specific gravity is typically expressed using three significant figures. For sands, G_s is often assumed to be 2.65 because this is the specific gravity of quartz. Since the mineralogy of clay is more variable, G_s for clay is more variable, and is often assumed to be somewhere between 2.70 and 2.80 depending on mineralogy.

3.4. EQUIPMENT AND MATERIALS

The following equipment and materials are required for specific gravity of soil solids measurements:

- Oven-dried soil sample;
- scale capable of measuring to the nearest 0.01 g;
- 500-ml etched flask;
- distilled or demineralized water;
- squeeze bottle;
- thermometer capable of reading to the nearest 0.5° C;
- funnel;
- stopper and tubing for connecting flask to vacuum supply; and
- vacuum supply capable of achieving a gauge vacuum of 660 mm Hg (12.8 psi).

Figure 3.1 is a photograph of the flask along with the stopper and tubing.

Fig. 3.1—Etched flask along with stopper and tubing for connecting to vacuum source.

3.5.   PROCEDURE[1]

The procedure presented herein is consistent with ASTM D854 Test Method A, where an oven-dried specimen of soil is used. The specific gravity calculation is based on three measurements:

1) Mass of the flask filled with distilled water to the etch mark, M_a;
2) mass of the flask filled with water and soil to the etch mark, M_b; and
3) mass of the dry soil, M_o.

Specific gravity of soil solids, G_s, is calculated based on these three parameters:

$$G_s = \frac{M_o}{M_o + (M_a - M_b)}.$$ (3.1)

Since the density of water is temperature-dependent, a temperature correction factor, K, may be applied to report G_s at a standard temperature of 20°C. The temperature-corrected G_s, G_{s20}, is expressed as:

$G_{s20} = G_s K.$ (3.2)

---

[1] Don't forget to visit www.wiley.com/college/kalinski to view the lab demo!

Table 3.1—Temperature correction factor, K, for reporting G_{s20}.

Temperature (^{o}C)	Correction Factor K
17	1.0006
18	1.0004
19	1.0002
20	1.0000
21	0.9998
22	0.9996
23	0.9993
24	0.9991

The procedure for performing the specific gravity measurement is as follows:

1) Weigh approximately 60 g of dry soil to obtain M_o.
2) Fill the flask to the etch line with distilled or demineralized water to obtain M_a.
3) Pour half of the water out of the flask and place the soil in the flask with a funnel.
4) Wash the soil down the inside neck of the flask.
5) Connect the flask to the vacuum source with the hose and stopper and apply vacuum for 30 minutes, occasionally agitating the mixture.
6) Fill the flask to the etch line with distilled water and weigh it to obtain M_b.
7) Record the water temperature in the flask and use Table 3.1 to obtain K.

3.6   EXPECTED RESULTS

Specific gravity of soil solids is controlled by soil mineralogy. In coarse-grained soils such as sands and gravels, where the mineralogy is dominated by quartz and feldspar, G_s is typically around 2.65. In fine-grained soils, G_s is more variable due to the presence of clay minerals, and may range from 2.70-2.85.

3.7.   LIKELY SOURCES OF ERROR

When measuring the specific gravity, the most likely source of error is inadequate de-airing of the soil mixture, which leads to an underestimate for G_s. According to ASTM D854, oven-dried clay specimens may require 2-4 hours of applied vacuum for adequate de-airing. However, for the purposes of demonstration in this lab, and to accommodate the typical three-hour laboratory class time, a de-airing time of 30 minutes is recommended. It is also recommended that a coarse-grained soil be used to improve the accuracy of the measurement given the short de-airing period.

3.8.   ADDITIONAL CONSIDERATIONS

In the absence of laboratory testing, G_s is often assumed based on the predominant mineralogy of the soil. However, certain types of soils, including organic soils, gypsum, and fly ash, possess values of G_s that are significantly less than the range of 2.65-2.85 often assumed by practicing engineers. Therefore, it is particularly important when dealing with such soils to measure G_s rather than assuming a value.

Finally, ASTM D854 includes criteria for assessing the acceptability of test results using this method. Assuming that all of the tests are performed by the same laboratory technician, G_s for two separate tests of the same material should be within 0.06 of each other to be considered acceptable.

3.9.   SUGGESTED EXERCISES

1)   Measure the specific gravity of the dry soil specimen supplied by the laboratory instructor using the Specific Gravity of Soil Solids Laboratory Data Sheet at the end of the chapter (additional data sheets can be found on the CD-ROM that accompanies this manual).

2)   If you did not adequately de-air your specific gravity specimen such that bubbles remained, would you overestimate or underestimate specific gravity? Why?

3)   If you were not able to perform a specific gravity test and had to estimate specific gravity for a sand and a clay, what values would you use?

SPECIFIC GRAVITY OF SOIL SOLIDS (ASTM D854)
LABORATORY DATA SHEET

I. GENERAL INFORMATION

Tested by:	Date tested:
Lab partners/organization:	
Client:	Project:
Boring no.:	Recovery depth:
Recovery date:	Recovery method:
Soil description:	

II. TEST DETAILS

Vacuum level:	Duration vacuum applied:
Flask volume:	
Scale type/precision/serial no.:	
Notes, observations, and deviations from ASTM D854 test standard:	

III. MEASUREMENTS AND CALCULATIONS

Test ID			
Mass of flask filled with water (M_a)			
Mass of flask filled with soil and water (M_b)			
Mass of dry soil (M_o)			
Specific gravity of soil solids (G_s)			
Water temperature			
Correction factor (K)			
Specific gravity of soil solids at 20°C (G_{s20})			

IV. EQUATION AND CALCULATION SPACE

$$G_s = \frac{M_o}{M_o + (M_a - M_b)}$$

$$G_{s20} = G_s K$$

4.   LIQUID AND PLASTIC LIMIT TESTING

4.1.   APPLICABLE ASTM STANDARD

- ASTM D4318: Standard Test Methods for Liquid Limit, Plastic Limit, and Plasticity Index of Soils

4.2.   PURPOSE OF MEASUREMENT

The liquid limit and plastic limit tests provide information regarding the effect of water content (w) on the mechanical properties of soil. Specifically, the effects of water content on volume change and soil consistency are addressed. The results of this test are used to classify soil in accordance with ASTM D2487, and to estimate the swell potential of soil.

4.3.   DEFINITIONS AND THEORY

The liquid and plastic limit are water contents at which the mechanical properties of soil changes. They are applicable to fine-grained soils, and are performed on soil fractions that pass the #40 (0.425-mm) sieve. Plastic limit (PL) and liquid limit (LL) are depicted in Fig. 4.1. The difference between the PL and the LL is defined as the plasticity index (PI):

$$PI = LL - PL. \tag{4.1}$$

In Fig. 4.1, the volume of fine-grained soil increases with increasing w. This indicates that PI is an indicator of the swell potential of a cohesive soil. Certain clay minerals, including bentonite, montmorillonite, and smectite, have a high cation exchange capacity, so their ability to hold water molecules and electrically bind them to their surface is greater. Therefore, they can exist in a plastic state over a relatively wide range of w and soil volume, and have a high swell potential.

A third value called the shrinkage limit (SL) is also depicted in Fig. 4.1. Shrinkage limit is the water content at which the volume of soil begins to change as a result of a change in w. The three parameters (SL, PL, and LL) are collectively referred to as the Atterberg limits. Shrinkage limit is measured using a separate standard, ASTM D427. However, shrinkage limit is not commonly specified in earthwork construction, and laboratory shrinkage limit testing includes the handling of mercury, which is not desirable for health and safety purposes. Therefore, the scope of this laboratory includes only plastic limit and liquid limit testing.

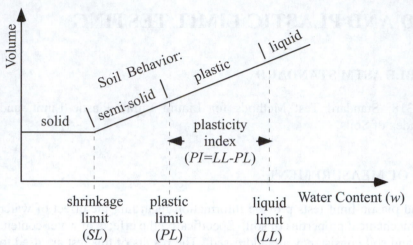

Fig. 4.1—Relationship between volume and water content in fine-grained soil.

4.4. EQUIPMENT AND MATERIALS

4.4.1. Liquid Limit Test

The following equipment and materials are required for liquid limit testing:

- Fine-grained soil;
- #40 sieve (0.425-mm opening);
- distilled or demineralized water;
- scale capable of measuring to the nearest 0.01 g;
- ceramic soil mixing bowl;
- soil drying oven set at $110° \pm 5°$ C;
- frosting knife;
- liquid limit device;
- grooving tool;
- 3 soil moisture containers; and
- permanent marker for labeling soil moisture containers.

4.4.2. Plastic Limit Test

The following equipment and materials required for plastic limit testing:

- Fine-grained soil;
- #40 sieve (0.425-mm opening);
- distilled or demineralized water;
- scale capable of measuring to the nearest 0.01 g;
- ceramic soil mixing bowl;
- soil drying oven set at $110° \pm 5°$ C;
- 0.125-in. diameter metal rod;
- frosted glass plate;

- 3 soil moisture containers; and
- permanent marker for labeling soil moisture containers;

4.5. PROCEDURE[1]

4.5.1 Liquid Limit Testing

The liquid limit is defined as the water content at which the soil starts to act as a liquid. To derive liquid limit, the following procedure, described as the Multipoint Method (Method A) in ASTM D4318, is described:

1) Pass the soil through a #40 sieve and use the fraction that passes the sieve.

2) Add distilled water to approximately 50 g of soil until it has the consistency of peanut butter or frosting.

3) Check that the drop height of the cup in the liquid limit device is 1.0 cm (Fig. 4.2), and adjust the apparatus as necessary. Most grooving tools have a tab with a dimension of exactly 1.0 cm that you can use.

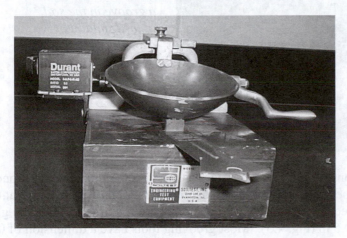

Fig. 4.2— Checking the drop height of the cup using the calibration tab on the grooving tool.

4) Spread a flat layer of soil in the cup with the frosting knife (Fig. 4.3).

---

[1] Don't forget to visit www.wiley.com/college/kalinski to view the lab demo!

Liquid and Plastic Limit Testing

Fig. 4.3—Spread a flat layer of soil in the liquid limit device cup prior to grooving.

5) Use the grooving tool to cut a groove in the soil (Fig. 4.4).

Fig. 4.4—Use the grooving tool to cut a groove in the soil in the liquid limit cup.

6) Turn the crank on the liquid limit device at a rate of 2 cranks per second and closely observe the groove. For each crank, the cup will drop from a height of 1.0 cm. Count and record the number of cranks that are required to close the groove over a length of 0.5 in (Fig. 4.5). Most grooving tools have a dimension of 0.5 in. that you can use.

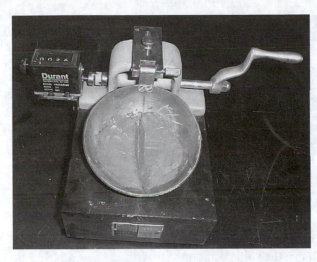

Fig. 4.5—The groove has closed over a length of 0.5 in.

7) Clean out the cup and repeat steps 4-6 until successive trials yield consistent results that are within a few cranks of each other, and record the average number of cranks for the soil.

8) Remove the soil from the cup, place it in a moisture container, and obtain its water content using the ASTM D2216 method described in Chapter 2.

The procedure outlined above will provide a data single point corresponding to a single number of cranks and single water content. Liquid limit is defined as the water content at which the groove closes at exactly 25 cranks. Most likely, it will require either more or less than 25 cranks to close the crack for the first test. To derive liquid limit using the multipoint method, the procedure is repeated at three different water contents, and the data are plotted on a semi-log graph of *w* versus number of cranks. The water content corresponding to 25 cranks (i.e. *LL*) is derived by interpolation. To obtain two additional points, add either water or soil to the original mixture (depending on *w* of the first point) and repeat the procedure.

4.5.2. Plastic Limit Test

The plastic limit is defined as the water content at which a 0.125-in. diameter rod of soil begins to crumble. It is measured using the following procedure:

1) Pass some soil through the #40 sieve and use the soil that passes the sieve;

2) Add some distilled water to make little mudballs that would stick to the wall if you threw them (DO NOT throw them).

3) Take a pea-sized mudball and roll it out onto the frosted plate to form a rod with a diameter of 0.125 in. Use the 0.125-in. diameter metal rod as a reference (Fig. 4.6). If the soil crumbles the first time, add more water and repeat.

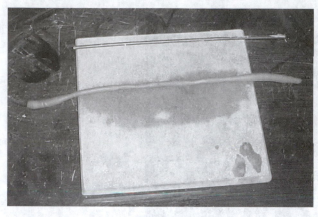

Fig. 4.6 – Rolling the soil to form a 0.125-in. diameter soil rod without crumbling.

4) If the rod doesn't crumble, pick it up and make another mudball in your hands. As you do this, you will dry the soil.

5) Repeat the process of making a rod, rolling up in your hands with a ball, making a rod, etc., until the soil crumbles while you are making the rod (Fig. 4.7). At this point, the water content of the soil is the *PL*. Quickly obtain its moist weight and place it in the oven for a moisture content reading in accordance with ASTM D2216 as described in Chapter 2.

Fig. 4.7—Soil rod crumbles at the plastic limit.

Repeat this entire procedure three times, and report an average value for the plastic limit.

4.6. EXPECTED RESULTS

Liquid limit typically ranges anywhere from 20% for silts to over 100% for high-plasticity clays. Plasticity index typically ranges anywhere from near 0% (i.e.; a non-plastic soil) for silts to over 50% for high-plasticity clays.

4.7. LIKELY SOURCES OF ERROR

Considering the seemingly archaic and empirical nature of these tests, one will find that the results obtained, particularly when plotting the three data points to obtain *LL*, are quite reliable. One likely source of error in performing these tests is in obtaining accurate water content measurements for the plastic limit test. Since the volume of soil used for the moisture content measurement is very small, significant moisture loss can occur while obtaining the moist weight of the soil specimen. The best way to minimize this error is to obtain the moist weight of the soil rod as quickly as possible after it crumbles.

4.8. ADDITIONAL CONSIDERATIONS

Plasticity index is a qualitative measure of the swell potential of soil. Clays with high cation exchange capacity, including bentonite, montmorillonite, and smectite, have high swell potentials. General guidelines for swell potential are summarized in Table 4.1.

The method described herein for liquid limit testing, Method A, relies on the use of three or more points and interpolation between points to derive the liquid limit. However, an alternative One-Point Method (Method B) is also described in ASTM D4318. With this method, one point with a moisture content w^n and corresponding number of cranks N is used to calculate *LL* with the following equation:

$$LL = w^n \left(\frac{N}{25} \right)^{0.121}.$$
(4.2)

Finally, ASTM D4318 includes criteria for assessing the acceptability of test results. Assuming that all of the tests are performed by the same laboratory technician, *LL* and *PL* for two separate tests of the same material should be within 2.4% and 2.6% of each other, respectively, to be considered acceptable.

Table 4.1—Ranges in LL and PI for typical fine-grained soil.

USCS Soil Type[1]	Common Mineralogy	Swell Potential	*LL*	*PI*
ML, CL	Kaolinite	Low	<50%	<25%
CL	Illite	Moderate	50-60%	25-35%
CH	Bentonite, Montmorillonite, Smectite	High	>60%	>35%

[1]see Chapter 6

4.9.   SUGGESTED EXERCISES

1)   Measure the liquid limit of the fine-grained soil provided in class using the Liquid Limit Data Sheet at the end of the chapter (additional data sheets can be found on the CD-ROM that accompanies this manual).

2)   Measure the plastic limit of the fine-grained soil provided in class using the attached Plastic Limit Data Sheet at the end of the chapter (additional data sheets can be found on the CD-ROM that accompanies this manual).

3)   Calculate the plasticity index of the fine-grained soil provided in class.

4)   Do these soils possess a high, moderate, or low swell potential?

5)   To measure the liquid limit, there are two methods described in ASTM D4318: Method A and Method B? Which method did you use? Briefly describe the method that you did not use.

LIQUID LIMIT (ASTM D4318)
LABORATORY DATA SHEET

I. GENERAL INFORMATION

Tested by:	Date tested:
Lab partners/organization:	
Client:	Project:
Boring no.:	Recovery depth:
Recovery date:	Recovery method:
Soil description:	

II. TEST DETAILS

Oven temperature:	Drying time:
Scale type/precision/serial no.:	
Notes, observations, and deviations from ASTM D4318 test standard:	

III. MEASUREMENTS AND CALCULATIONS

Trial Number	1	2	3
Container ID			
Mass of container (M_c)			
Mass of moist soil + container (M_1)			
Mass of dry soil + container (M_2)			
Mass of moisture (M_w)			
Mass of dry soil (M_s)			
Moisture Content (w)			
Number of Cranks			
Liquid Limit (LL)			
Corresponding Plastic Limit (PL)			
Plasticity Index (PI)			

IV. EQUATION AND CALCULATION SPACE

$$w = \frac{M_1 - M_2}{M_2 - M_c} \times 100\%$$

$$PI = LL - PL$$

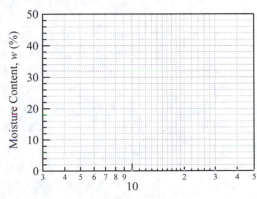

Number of Cranks

PLASTIC LIMIT (ASTM D4318)
LABORATORY DATA SHEET

I. GENERAL INFORMATION

Tested by:	Date tested:
Lab partners/organization:	
Client:	Project:
Boring no.:	Recovery depth:
Recovery date:	Recovery method:
Soil description:	

II. TEST DETAILS

Oven temperature:	Drying time:
Scale type/precision/serial no.:	
Notes, observations, and deviations from ASTM D4318 test standard:	

III. MEASUREMENTS AND CALCULATIONS

Trial Number	1	2	3
Container ID			
Mass of container (M_c)			
Mass of moist soil + container (M_1)			
Mass of dry soil + container (M_2)			
Mass of moisture (M_w)			
Mass of dry soil (M_s)			
Moisture Content (w)			
Average Plastic Limit (PL)			
Corresponding Liquid Limit (LL)			
Plasticity Index (PI)			

IV. EQUATION AND CALCULATION SPACE

$$w = \frac{M_1 - M_2}{M_2 - M_c} \times 100\%$$

$$PI = LL - PL$$

5.   ANALYSIS OF GRAIN SIZE DISTRIBUTION

5.1.   APPLICABLE ASTM STANDARDS

- ASTM D422: Standard Test Method for Particle-Size Analysis of Soils

- ASTM D1140: Standard Test Method for Amount of Material in Soils Finer Than the No. 200 (75-μm) Sieve

5.2.   PURPOSE OF MEASUREMENT

The purpose of these tests is to determine the grain size distribution (i.e.; grain size versus percent by weight) of soil, and to determine the percentage of fines (i.e.; material passing the No. 200 sieve) in soil. This information is used to classify the soil in accordance with the Unified Soil Classification System (USCS).

5.3.   DEFINITIONS AND THEORY

5.3.1. Mechanical Sieving

Soil consists of individual particles, or grains. Grain size refers to the size of an opening in a square mesh through which a grain will pass. Since all of the grains in a mass of soil are not the same size, it is convenient to quantify grain size in terms of a gradation curve. A gradation curve contains points corresponding to a particular grain size, and a corresponding percent (by weight) of the soil grains that are smaller than that grain size. In the example shown in Fig. 5.1, 30% of the soil grains are smaller than 0.18 mm.

To perform grain size analysis of a dry granular soil (sand or gravel), mechanical sieving is used, and the soil is passed through a stack of sieves. Any number of sieves can be used, but the size of the stack is typically limited to six sieves. The coarsest sieve is at the top of the stack, followed by increasingly finer sieves below. A pan is placed below the bottom sieve to collect the soil that passes the finest sieve. By weighing the fraction retained by each sieve, points on the gradation curve can be calculated.

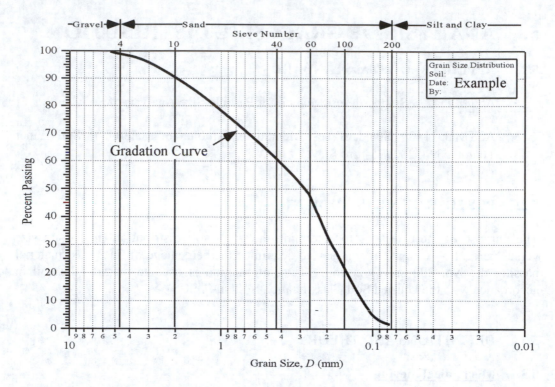

Fig. 5.1—Typical gradation curve.

5.3.2. Hydrometer Analysis

In some instances, the gradation curve cannot be reliably quantified at smaller grain sizes (less than a millimeter) using sieves because the smaller clay particles in soil form clods and cannot pass through the screens individually. However, this portion of the gradation curve can be defined using a hydrometer analysis. A hydrometer is a bulb that is heavily weighted at the bottom, with a graduated neck on the top (Fig. 5.2).

When the hydrometer is placed in a fluid, it floats like a fishing bobber. The density of the fluid affects the buoyancy of the hydrometer. Denser fluids allow the hydrometer to be more buoyant and float higher. To perform a hydrometer analysis, soil is mixed with water and sodium hexametaphosphate (a dispersing agent) to create a slurry of dispersed soil particles. The soil particles are initially suspended in the liquid mixture, but settle over time. Larger particles settle faster in accordance with Stokes' Law, which states the diameter of a spherical particle is proportional to the square root of its settling velocity. As smaller and smaller particles settle past the center of mass of the hydrometer with the passage of time, the density of the slurry affecting the buoyancy of the hydrometer decreases, and the hydrometer floats lower and lower in the slurry. Information regarding how low the hydrometer floats in the slurry is recorded as a function of time, and this information is used to calculate points of grain size versus percent passing for the gradation curve.

5.3.3. <u>Wet Sieving</u>

The amount of fines (i.e.; grain size smaller than 75 μm corresponding to a #200 sieve) in soil plays an important role in soil behavior and classification. In some instances, information regarding fines content may be desired without the need to fully define the entire gradation curve. Mechanical sieving may produce erroneous results because the smaller particles form clods, while a hydrometer analysis may be too rigorous. A simpler alternative may be to perform wet sieving of the soil. In wet sieving, soil is combined with water and sodium hexametaphosphate to disperse the flocculated clay particles. Flocculation occurs because fine-grained soil particles are platy, and possess negative charges on their faces and positive charges on their edges. As a result, the particles are attracted to one another in an edge-to-face manner to form clods. Sodium hexametaphosphate neutralizes the surface charges on the clay particles, which disperses the particles and allows them to individually pass through the #200 sieve. The slurry is passed through a #200 sieve, which yields a more accurate estimate for the percentage of fines in the soil.

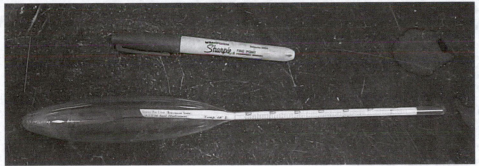

Fig. 5.2—Photograph of a hydrometer (pen shown for scale).

5.4. EQUIPMENT AND MATERIALS

5.4.1. <u>Mechanical Sieve Analysis</u>

The following equipment and materials are required for performing a mechanical sieve analysis of soil to partially define the gradation curve:

- Oven-dried soil;
- sieve stack consisting of, from top to bottom;
  - lid;
  - #4 sieve (4.75 mm opening);
  - #10 sieve (2.00 mm opening);
  - #40 sieve (0.425-mm opening); and
  - Pan.
- scale capable of measuring to the nearest 0.01 g;
- mechanical shaker (optional); and
- timing device capable of reading to the nearest second.

Analysis of Grain Size Distribution

5.4.2. Hydrometer Analysis

The following equipment and materials are required for performing a hydrometer analysis of soil to partially define the gradation curve:

- Oven-dried soil passing the #40 sieve;
- scale capable of measuring to the nearest 0.01 g;
- distilled or demineralized water;
- 152H type hydrometer;
- 40 g/l sodium hexametaphosphate solution;
- 250-ml beaker;
- ASTM D422-specified stirring device and dispersion cup;
- 1000-ml etched graduated cylinder;
- 1000-ml graduated cylinder;
- rubber stopper for the etched graduated cylinder;
- timing device capable of reading to the nearest second;
- thermometer capable of reading to the nearest 0.5° C; and
- squeeze bottle.

5.4.3. Wet Sieve Analysis

The following equipment and materials are required for performing a wet sieve analysis of soil to measure the fines content:

- Oven-dried soil;
- scale capable of measuring to the nearest 0.01 g;
- squeeze bottle;
- deep (greater than 6 in.) #200 sieve with reinforcement to prevent screen damage;
- large oven-safe mixing bowl;
- 40 g/l sodium hexametaphosphate solution;
- sink with running tap water; and
- large soil drying oven set at $110^\circ \pm 5^\circ$ C.

5.5. PROCEDURE[1]

5.5.1. Mechanical Sieve Analysis (ASTM D422)

Record your measurements and calculations on the Grain Size Analysis Data Sheet using the following procedure:

1) Place approximately 750 g of soil (M_{total}) in the top of the sieve stack.

---

[1] Don't forget to visit www.wiley.com/college/kalinski to view the lab demo!

2) Shake the sieve stack manually for 10 minutes while keeping the stack upright. Alternatively, you may place the sieve stack in a mechanical shaker and shake for 5 minutes. Dust masks and ear protection are recommended for this step.

3) The material in the pan passed the #40 sieve. Measure and record its net mass ($M_{-\#40}$). Divide $M_{-\#40}$ by M_{total} to obtain the percentage of soil that passed the #40 sieve ($P_{-\#40}$).

4) Set the soil that passed the #40 sieve aside. This soil will be used for the hydrometer analysis.

5) Measure the mass of the soil directly on top of the #40 sieve and add this mass to $M_{-\#40}$. This sum represents the soil that passed the #10 sieve ($M_{-\#10}$). Divide $M_{-\#10}$ by M_{total} to obtain the percentage of soil that passed the #10 sieve ($P_{-\#10}$).

6) Measure the mass of the soil directly on top of the #10 sieve and add this mass to $M_{-\#10}$. This sum represents the soil that passed the #4 sieve ($M_{-\#4}$). Divide $M_{-\#4}$ by M_{total} to obtain the percentage of soil that passed the #4 sieve ($P_{-\#4}$).

7) Measure the mass of the soil directly on top of the #4 sieve. This mass represents the soil retained by the #4 sieve ($M_{+\#4}$).

8) Add $M_{-\#4}$ and $M_{+\#4}$ to calculate the total mass of soil after sieving, M_{total}'. Record this mass on the Mechanical Sieve Data Sheet, along the with percent soil loss:

$$\% \, loss = \frac{M_{total} - M_{total}'}{M_{total}} \, x \, 100\% \, . \tag{5.1}$$

5.5.2. <u>Hydrometer Analysis (ASTM D422)</u>

The material that passed the #40 sieve during the mechanical sieve analysis is used to perform the hydrometer analysis. Record your measurements and calculations on the Grain Size Analysis Data Sheet using the following procedure:

1) Combine approximately 50.0 g (M_d) of the soil that passed the #40 sieve with 125 ml of the sodium hexametaphospahte solution in a 250-ml glass beaker. Allow the mixture to soak for at least 16 hours in accordance with ASTM D422 procedures (NOTE: a 30-minute soaking period may be used for demonstration and educational purposes).

2) Transfer all of the mixture to an ASTM D422-specified dispersion cup (Fig. 5.3). Use a squeeze bottle of distilled water to wash all of the soil solids from the inside of the beaker into the dispersion cup. After transfer, the dispersion cup should be more than half full of mixture.

Fig. 5.3—ASTM D422-specified dispersion cup (2 cups shown with pen for scale; note the bafflers inside the cup).

3) Stir the mixture using an ASTM D422-specified stirring device at a rate of 10,000 rpm for one minute (Fig. 5.4).

Fig. 5.4—ASTM D422-specified stirring device (shown with dispersion cup).

4) Pour the slurry into a 1000-ml etched cylinder and fill with distilled water to just below the etch mark. Use a squeeze bottle of distilled water to wash all of the slurry from the cup into the cylinder.

5) Using a rubber stopper, mix the cylinder by turning it upside down and back at a rate of 1 turn per second for 1 minute (NOTE: turning the cylinder upside down and back counts as two turns).

6) Set the cylinder down and start the timer immediately. Using the squeeze bottle, wash the remaining soil off the stopper and lip of the cylinder down into the cylinder, and fill the cylinder to the etch mark with distilled water.

7) Take your first hydrometer reading at 2 minutes, with subsequent readings at 5, 15, 30, 60, 250, and 1440 minutes (the 250- and 1440-minute readings may be replaced with 90- and 120-minute readings for educational purposes). The hydrometer reading, R, is read off the neck of the hydrometer at the top of the meniscus (Fig. 5.5). Record the time, t, in minutes.

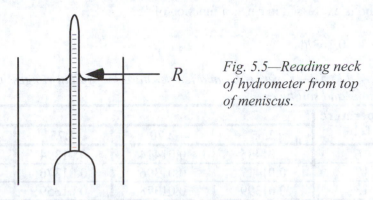

R

Fig. 5.5—Reading neck of hydrometer from top of meniscus.

8) Remove the hydrometer after each reading, and place it in a 1000-ml cylinder filled with distilled water between readings. Spin the hydrometer while it is in this cylinder to remove adhered soil particles (Fig. 5.6).

Fig. 5.6—Hydrometer placed in second cylinder between readings.

9) Record the water temperature in the cylinder containing the soil slurry and estimate G_s. If distilled water at room temperature is used for the test, and the room is kept at a constant temperature, a single water temperature reading should suffice.

10) At a given time t, particles larger than D have settled past the center of mass of the hydrometer and no longer affect its buoyancy. Use Stoke's Law to calculate the particle diameter, D, in mm, corresponding to t in minutes:

$$D = K\sqrt{L/t}.$$ (5.2)

In Eqn. 5.2, K is a function of temperature and G_s, which both affect the density of the slurry (Table 5.1). The parameter L represents the distance between the center of mass of the hydrometer and the point where the hydrometer is read (Fig. 5.4), and is expressed in cm as a function of R:

$$L = 16.3 - 0.163R.$$ (5.3)

Table 5.1—K versus G_s and temperature for typical ranges in laboratory conditions and soil types.

Temperature (°C)	G_s			
	2.65	2.70	2.75	2.80
16	0.01435	0.01414	0.01394	0.01374
17	0.01417	0.01396	0.01376	0.01356
18	0.01399	0.01378	0.01359	0.01339
19	0.01382	0.01361	0.01342	0.01323
20	0.01365	0.01344	0.01325	0.01307
21	0.01348	0.01328	0.01309	0.01291
22	0.01332	0.01312	0.01294	0.01276
23	0.01317	0.01297	0.01279	0.01261
24	0.01301	0.01282	0.01264	0.01246
25	0.01286	0.01267	0.01249	0.01232

As shown in Fig. 5.7, the hydrometer floats high in the slurry at the start of the test, but sinks with the passage of time as soil solids settle and the density of the slurry decreases. The total change in R during the test is a function of G_s, water temperature, and the soil concentration.

11) For each measurement, use the following equation to calculate the percent passing, P', corresponding to D:

$$P' = \frac{(R-b)a}{M_d} \; x \; 100\%.$$ (5.5)

In this equation, M_d is the oven dried mass of the soil in the slurry (approximately 50.0 g). Since the hydrometer is calibrated for $G_s = 2.65$, the correction factor a is used to account for deviations in G_s from 2.65. The "composite" correction factor b is used to account for the effects of i) sodium hexametaphosphate on slurry density, ii) deviations from the hydrometer calibration temperature of 20°C, and

iii) reading from the top of the meniscus instead of the bottom. Values for *a* and *b* are given Table 5.2.

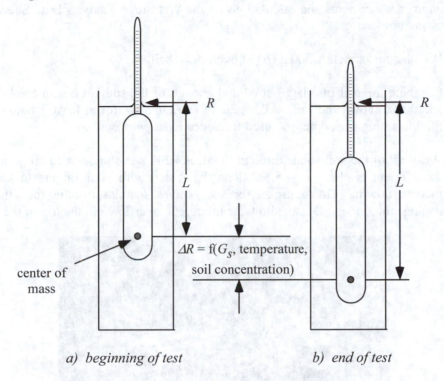

a) beginning of test   b) end of test

Fig. 5.7—Appearance of hydrometer at beginning of test (t = 0) and end of test (t → ∞).

Table 5.2 —Correction factors a and b for calculation of P'.

a		b	
G_s	a	Temp. (^{0}C)	b
2.50	1.03	17	5.9
2.55	1.02	18	5.6
2.60	1.01	19	5.3
2.65	1.00	20	5.0
2.70	0.99	21	4.7
2.75	0.98	22	4.4
2.80	0.97	23	4.1
2.85	0.96	24	3.8

12) Since the values for *P'* from the hydrometer test were derived using the fraction of the soil passing the #40 sieve, they must be multiplied by $P_{-\#40}$ for plotting with the points derived by mechanical sieving.

5.5.3. Wet Sieve Analysis (ASTM D1140)

Record your measurements and calculations on the Wet Sieve Analysis Data Sheet using the following procedure:

1) Weigh approximately 500 g (B) of oven-dried soil.

2) Combine the soil in a large bowl and enough of the sodium hexametaphosphate solution to cover the soil. Allow the mixture to soak for at least 2 hours (a 30-minute sitting period may be used for educational purposes).

3) Wash all of the soil solids through the deep #200 sieve under a running tap until the effluent is clear (Fig. 5.8). Rub the screen with your fingers to keep the mixture flowing. Do not use any brushes, knives, spatulas, or other tools that may damage the screen. Do not allow the mixture to overflow out the top of the sieve.

Fig. 5.8—Washing the soil through a #200 sieve.

4) Wash all of the solids retained in the sieve back into the mixing bowl using tap water and a squeeze bottle. It is alright to have a large amount of water in the bowl, provided it is not spilling over the side of the bowl.

5) Place the bowl in a large drying oven and let dry overnight.

6) Calculate the net dry mass of the soil retained by the #200 sieve, C.

7) Calculate the percent fines in the soil, A:

$$A = \frac{B-C}{B} \; x \; 100\%.$$

(5.6)

5.6. LIKELY SOURCES OF ERROR

Likely sources of error for the grain size analysis tests include:

- *Holes in the sieves.* Sieves should be inspected and repaired as needed prior to sieving.

- *Significant soil loss during sieving.* Soil may be lost by escaping out the sides of the sieves, or becoming lodged in the screens during sieving. Soil particles from previous sieving activities may also become dislodged during sieving, leading to a final total mass that is greater than the initial total mass. Sieves should be cleaned with a sieve brush prior to sieving, and the *% loss* calculated after sieving should be less than a few percent.

- *Inadequate dispersion of clay particles.* Soil is combined with sodium hexametaphosphate solution during hydrometer analysis and wet sieving to disperse particles. To achieve adequate dispersion, ASTM D422 and D1140 state that the soil should be allowed to soak in the solution for at least 16 hours and 2 hours, respectively. Herein, a 30-minute soaking period is recommended for each test given the time constraints of a typical undergraduate soil mechanics laboratory.

- *Undermixing or overmixing of soil slurry prior to hydrometer testing.* Prior to hydrometer testing, the soil-sodium hexametaphosphate slurry is mixed for one minute. Mixing for less than one minute may result in incomplete dispersion, while mixing for more than one minute may result in soil particle breakage, which will affect grain size distribution.

- *Leaving the hydrometer in the slurry between readings.* The hydrometer must not be left in the soil slurry between readings. If left in the slurry, soil particles will begin to adhere to the hydrometer and affect its buoyancy.

5.7. ADDITIONAL CONSIDERATIONS

Of the three tests described, the dry sieving and wet sieving are most commonly used and most valuable with respect to soil classification using the Unified Soil Classification System (USCS). USCS soil classification does not make a distinction between particle sizes for particles smaller than 75 μm, while the hydrometer test primarily gives information regarding gradation of soil with sizes less than 75 μm. One unique use for hydrometer test results is in the measurement of soil activity. Soil activity is the slope of a curve of *PI* versus percent passing 2 μm. The greater the activity, the more susceptible the soil is to shrinking and swelling. However, geotechnical engineers in the United States more commonly use *PI* as a measure of swell potential.

The sieve sizes recommended for mechanical sieving in this laboratory exercise (#4, #10, and #40) are based on the assumption that the soil contains little or no gravel-sized particles (i.e.; particles retained by the #4 sieve). However, if a gravelly soil is analyzed, sieves with sizes up to 3 in. can be added to the stack to define the coarser portion of the gradation curve.

According to ASTM D422, hydrometer testing can be performed on material passing the #10, #40, or #200 sieve. For this exercise, the #40 sieve was selected based on engineering judgment. For the cutoff sieve size selected, the material retained by that sieve should exist as individual particles rather than clods. If you observed clods on top of the #40 sieve after mechanical sieving, you would probably want to perform the hydrometer test on material that passed the #10 sieve. Alternatively, if your soil had little or no clay and you did not observe any clods for particles sizes down to those retained by the #200 sieve, you would probably want to perform the hydrometer test on material that passed the #200 sieve.

For mechanical sieving, the minimum mass of the fraction retained on the cutoff sieve (either #10, #40, or #200) increases with increasing maximum particle size (Fig. 5.9), while the minimum mass of the fraction passing the cutoff sieve is 115 g and 65 g for sandy and silty/clayey soils, respectively. For wet sieving, the minimum mass of the test specimen also increases with increasing maximum particle size (Fig. 5.10).

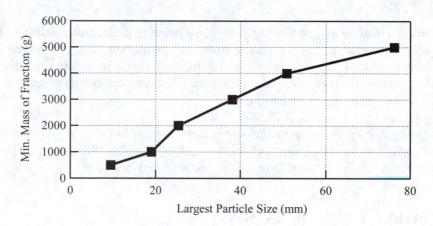

Fig. 5.9—Minimum mass of fraction to be used for mechanical sieving versus largest particle size in specimen.

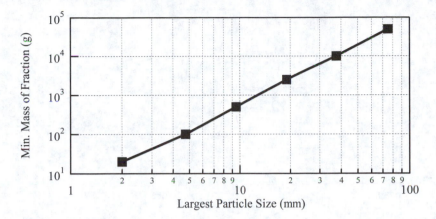

Fig. 5.10—Minimum mass of test specimen to be used for wet sieving versus largest particle size in specimen.

5.8. SUGGESTED EXERCISES

1) Perform a grain size analysis, including mechanical sieve and hydrometer analysis, on the soil supplied by the instructor. Use the Grain Size Analysis Data Sheet and Gradation Curve Form at the end of the chapter (additional data sheets can be found on the CD-ROM that accompanies this manual).

2) Calculate the weight of fines of the soil supplied by the instructor using the wet sieving method. Use the Weight of Fines Analysis Data Sheet at the end of the chapter (additional data sheets can be found on the CD-ROM that accompanies this manual).

3) What is flocculation and what does it have to do with wet sieving and hydrometer testing?

GRAIN SIZE ANALYSIS – DRY SIEVE MEASUREMENT (ASTM D422)
LABORATORY DATA SHEET

I. GENERAL INFORMATION

Tested by:	Date tested:
Lab partners/organization:	
Client:	Project:
Boring no.:	Recovery depth:
Recovery date:	Recovery method:
Soil description:	

II. TEST DETAILS

Sieve shaking method/duration:
Total sample mass before sieving (M_{total}):
Total sample mass after sieving (M_{total}'):
Percent soil loss during sieving (% $loss$):

III. MEASUREMENTS AND CALCULATIONS

Sieve Number	Sieve Opening, D (mm)	Cumulative Mass of Soil Passing (g)	Cumulative Mass of Soil Retained (g)	Percent Passing, P (%)
4	4.75	$M_{-\#4}=$	$M_{+\#4}=$	$P_{-\#4}=$
10	2.00	$M_{-\#10}=$	--	$P_{-\#10}=$
40	0.425	$M_{-\#40}=$	--	$P_{-\#40}=$

IV. EQUATION AND CALCULATION SPACE

$$\% \ loss = \frac{M_{total} - M_{total}'}{M_{total}} \ x \ 100\%$$

GRAIN SIZE ANALYSIS – HYDROMETER MEASUREMENT (ASTM D422)
LABORATORY DATA SHEET

I. GENERAL INFORMATION

Tested by:	Date tested:
Lab partners/organization:	
Client:	Project:
Boring no.:	Recovery depth:
Recovery date:	Recovery method:
Soil description:	

II. TEST DETAILS

Hydrometer manufacturer/serial no.:		
Mixer manufacturer/serial no.:		
Scale type/serial no./precision:		
Duration of initial soaking period:		
Concentration of sodium hexametaphosphate solution:		
Dry mass of soil used (M_d):		
Specific gravity of soil solids:	Temperature:	
K:	a:	b:
Notes, observations, and deviations from ASTM D422 test standard:		

III. MEASUREMENTS AND CALCULATIONS

Clock Time (hh:mm:ss)	t (min)	R	L (cm)	D (mm)	P' (%)	P (%)

IV. EQUATION AND CALCULATION SPACE

$$L = 16.3 - 0.163R \qquad D = K\sqrt{L/t}$$

$$P' = \frac{(R-b)a}{M_d} \times 100\% \qquad P = P'(P_{-\#40})$$

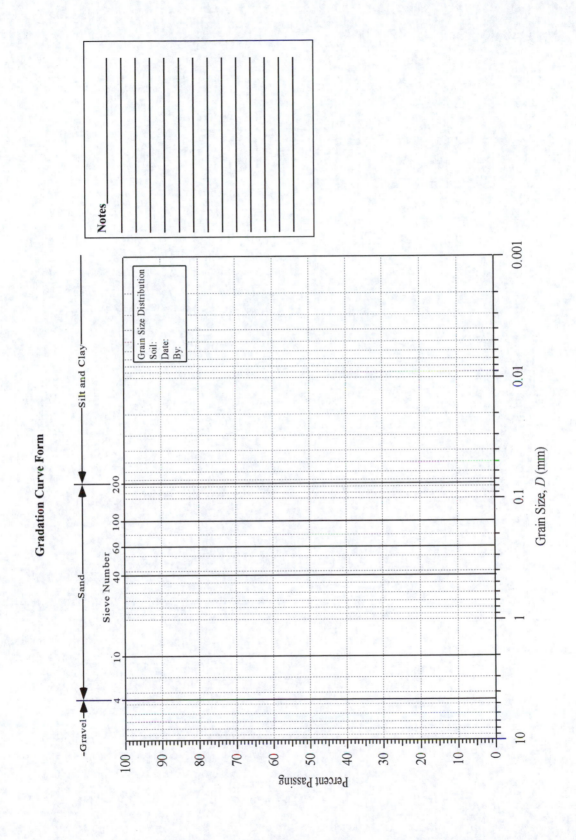

Gradation Curve Form

Notes

WET SIEVE ANALYSIS DATA SHEET (ASTM D1140)
LABORATORY DATA SHEET

I. GENERAL INFORMATION

Tested by:	Date tested:
Lab partners/organization:	
Client:	Project:
Boring no.:	Recovery depth:
Recovery date:	Recovery method:
Soil description:	

II. TEST DETAILS

Scale type/serial no./precision:	
Oven temperature:	Duration of oven drying:
Concentration of sodium hexametaphosphate solution:	
Duration of soaking period:	
Notes, observations, and deviations from ASTM D1140 test standard:	

III. MEASUREMENTS AND CALCULATIONS

Net dry mass of soil before sieving (*B*):
Net dry mass of soil retained by the #200 sieve (*C*):
Percent fines (*A*):

IV. EQUATION AND CALCULATION SPACE

$$A = \frac{B - C}{B} \, x \, 100\%$$

6.   LABORATORY CLASSIFICATION OF SOIL

6.1.   APPLICABLE ASTM STANDARD

- ASTM D2487:   Standard Practice for Classification of Soils for Engineering Purposes (Unified Soil Classification System)

6.2.   PURPOSE OF MEASUREMENT

Soil is classified by geotechnical engineers for engineering purposes in accordance with the Unified Soil Classification System (USCS).   Soils sharing a common USCS classification possess similar engineering properties, including strength, permeability, and compressibility, so the USCS is useful for specifying soil types to achieve a desired performance.

6.3.   DEFINITIONS AND THEORY

The USCS allows soil to be classified based on its engineering properties, including strength, permeability, and compressibility. To use the USCS, information regarding the liquid and plastic limits and gradation of the soil is required. Using the USCS, each soil is assigned a two-letter group symbol and a group name. The three basic soil types and the group symbols that fall under each soil type are:

- *Gravels:*   *GP, GW, GM, and GC,*
- *Sands:*   *SP, SW, SM, and SC, and*
- *Silts and Clays:*   *ML, CL, CH, MH, OH, and OL.*

Under the USCS, there is no direct distinction between silts and clays, although clay particles are smaller than silt particles and are mineralogically different than silt particles. Silts and clays are indirectly distinguished in the USCS through the use of liquid and plastic limits as described later. Although there are six group symbols listed under silts and clays, the last three symbols (MH, OH, and OL) are relatively uncommon.

Each group symbol has two letters. The first letter describes the soil type as follows:

- *G = gravel;*
- *S = sand;*
- *M = silt (muck);*
- *C = clay; and*
- *O = organic.*

The second letter is a modifier that provides additional description of the soil:

- *P = poorly graded;*
- *W = well graded;*
- *M = silty;*
- *C = clayey;*
- *L = low-plasticity (lean); and*
- *H = high-plasticity (fat).*

In addition to the group symbol, each soil is assigned a group name, which further modifies and describes the soil.

6.4.   EQUIPMENT AND MATERIALS

USCS classification can be performed using the instructions provided herein, but use of tables and charts; such as those that are published in ASTM D2487 and most undergraduate soil mechanics textbooks, may also facilitate the process.

6.5.   PROCEDURE

USCS soil classification is a methodical procedure that follows these steps:

1) ***Decide if the soil is fine-grained or coarse-grained.*** If more than 50% of the soil passes the #200 sieve, it is fine-grained. Otherwise, it is coarse-grained.

2a) ***For fine-grained soils,*** plot the *LL* and *PI* on the plasticity chart (Fig. 6.1). The point will fall in the quadrant corresponding to the USCS group symbol, which will most likely be either a silt (**ML**), lean clay (**CL**), or fat clay (**CH**).

2b) ***For coarse-grained soils,*** *determine if the soil is a sand or a gravel.* The material retained by the #200 sieve is referred to as the coarse fraction. If more than 50% of the coarse fraction passes the #4 sieve, the soil is a sand. Otherwise, it is a gravel.

3a)   ***For sands,*** *determine if it is a clean sand a dirty sand, or a dual classification.* If less than 5% of the soil passes the #200 sieve, it is a clean sand. If greater than 12% of the soil passes the #200 sieve, it is a dirty sand. If 5-12% pass the #200 sieve, it is a dual classification.

For clean sands, determine if it is well graded or poorly graded. Calculate the coefficient of uniformity, c_u, and the coefficient of curvature, c_c, on the gradation curve:

$$c_u = \frac{D_{60}}{D_{10}} \text{ and} \qquad\qquad\qquad (6.1)$$

$$c_c = \frac{(D_{30})^2}{D_{60}D_{10}}, \tag{6.2}$$

where D_{10}, D_{30}, and D_{60} are the grain sizes corresponding to 10%, 30%, and 60% passing, respectively. If $c_u > 6$ and $1 < c_c < 3$, the soil is a well-graded sand (**SW**). Otherwise, it is a poorly-graded sand (**SP**).

For dirty sand, determine if it is a silty sand or a clayey sand. Plot the *LL* and *PI* limits on the Plasticity Chart. If the point plots above the A-line, it is a clayey sand (**SC**). If it plots below the A-line, it is a silty sand (**SM**).

For dual classification: use the procedure for both clean sands and dirty sands to provide a four-letter dual classification, which may be a well-graded sand with silt (**SW-SM**), well-graded sand with clay (**SW-SC**), poorly-graded sand with silt (**SP-SM**), or poorly-graded sand with clay (**SP-SC**).

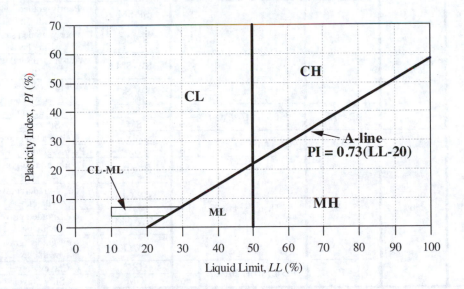

Fig. 6.1—Plasticity chart.

3b) *For gravels, determine if it is a clean gravel, a dirty gravel, or a dual classification.* If less than 5% of the soil passes the #200 sieve, it is a clean gravel. If greater than 12% of the soil passes the #200 sieve, it is a dirty gravel. If 5-12% pass the #200 sieve, it is a dual classification.

For clean gravel, determine if it is well graded or poorly graded. Calculate the coefficient of uniformity, c_u, and the coefficient of curvature, c_c, on the gradation curve. If $c_u > 4$ and $1 < c_c < 3$, the soil is a well-graded gravel (**GW**). Otherwise, it is a poorly-graded gravel (**GP**).

For dirty gravel, determine if it is a silty gravel or a clayey gravel. Plot the *LL* and *PI* limits on the Plasticity Chart. If the point plots above the A-line, it is a clayey gravel (**GC**). If it plots below the A-line, it is a silty gravel (**GM**).

For dual classification: use the procedure for both clean gravel and dirty gravel to provide a four-letter dual classification, which may be a well-graded gravel with silt (**GW-GM**), well-graded gravel with clay (**GW-GC**), poorly-graded gravel with silt (**GP-GM**), or poorly-graded gravel with clay (**GP-GC**).

The overall USCS procedure is represented in Fig. 6.2. To use this chart, start at the left side and work towards the right.

Coarse-Grained Soils

% passing #200	% of C.F. passing #4	% passing #200			USCS Symbol	USCS Name
<50%	>50%	0-5%	c_u>6 and 1<c_c<3?	yes	SW	Well-graded sand
				no	SP	Poorly-graded sand
		5-12%	Dual classification		SP-SM	Poorly-graded sand with silt
					SP-SC	Poorly-graded sand with clay
					SW-SM	Well-graded sand with silt
					SW-SC	Well-graded sand with clay
		12-50%	PI>0.73(LL-20)%?	yes	SC	Clayey sand
				no	SM	Silty sand
	<50%	0-5%	c_u>4 and 1<c_c<3?	yes	GW	Well-graded gravel
				no	GP	Poorly-graded gravel
		5-12%	Dual classification		GP-GM	Poorly-graded gravel with silt
					GP-GC	Poorly-graded gravel with clay
					GW-GM	Well-graded gravel with silt
					GW-GC	Well-graded gravel with clay
		12-50%	PI>0.73(LL-20)%?	yes	GC	Clayey gravel
				no	GM	Silty gravel

Fine-Grained Soils

% passing #200?	LL > 50%?	PI > 0.73(LL-20)%?	USGS Symbol	USCS Name
>50%	yes	yes	CH	Fat clay
		no	MH	Elastic silt
	no	yes	CL	Lean clay
		no	ML	Lean silt

Fig.6.2—USCS classification chart.

6.6.   LIKELY SOURCES OF ERROR

Error in soil classification is a result of error in the *LL* and *PI* or gradation tests, provided that the USCS has been used properly. The most common error in *LL* and *PI* testing is allowing the plasticity index specimens to sit too long before obtaining their moist weight. This error would result in underestimating *PL* and overestimating *PI*, and may result in erroneously classifying low-plasticity soils as high-plasticity soils. The most common error in gradation testing is underestimating the percent of fines in soil by relying on mechanical sieve analysis rather than wet sieve analysis to calculate fines content. This may result in erroneously classifying silty or clayey sands and gravels as clean sands and gravels.

6.7.   ADDITIONAL CONSIDERATIONS

As mentioned previously, soils sharing a common USCS group symbol possess similar engineering properties. Table 6.1 summarizes soil types that provide various performance.

Table 6.1 – USCS soil types and soil performance.

To achieve:	Use	Feature
Low permeability	ML, CL, CH	Fine-grained
High permeability	GP, SP	Poorly-graded
High strength	GW, SW	Well-graded
Low compressibility	GM, GP, GW	Gravelly

The procedure for classifying soil using the USCS is described herein. However, the American Association of State Highway Transportation Officials (AASHTO) has also developed a soil classification system that is extensively used for transportation-related earthworks. The reader should be aware of the AASHTO system. Information regarding use of the AASHTO system can be found in numerous other references.

As mentioned previously, USCS classification includes a two-letter group symbol, and a more descriptive group name. Details for determining the more descriptive group name are not given herein, but can be found in the ASTM D2487 standard.

6.8.   SUGGESTED EXERCISES

1)   Classify the soils in the table on the top of the following pate using the Unified Soil Classification System. Use the Gradation Curve Form at the end of the chapter to plot gradation curves as needed (additional data sheets can be found on the CD-ROM that accompanies this manual).

Use for Exercise No. 1

Soil #	Sieve Analysis, % finer than:							LL(%)	PI (%)
	#4 sieve (4.750 mm)	#10 sieve (2.000 mm)	#20 sieve (0.850 mm)	#40 sieve (0.425 mm)	#60 sieve (0.250 mm)	#140 sieve (0.106 mm)	#200 sieve (0.075 mm)		
1a	94	63	21	10	7	4	3	Non-plastic	Non-plastic
1b	98	86	50	28	18	12	10	Non-plastic	Non-plastic
1c	100	100	98	93	88	80	77	63	40
1d	100	100	100	99	95	88	86	45	28
1e	100	100	100	94	82	58	45	36	9

2) Classify the soils in the table below using the Unified Soil Classification System. Refer to the gradation curves on the following page and fill in all the blanks in the table below. Show all your work.

Soil #	LL (%)	PI (%)	USCS Classification
2a	N/A	N/A	
2b	55	30	
2c	Non-plastic	Non-plastic	
2d	30	25	
2e	40	10	

Use for Exercise No. 2

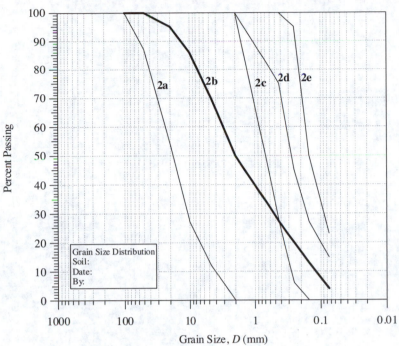

Laboratory Classification of Soil

Gradation Curve Form

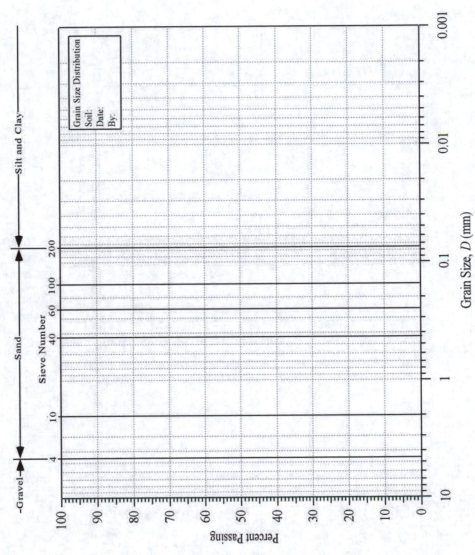

Notes _____

7. FIELD CLASSIFICATION OF SOIL

7.1. APPLICABLE ASTM STANDARDS

- ASTM D2488: Standard Practice for Description and Identification of Soils (Visual-Manual Procedure)

7.2. PURPOSE OF MEASUREMENT

Soil is classified for engineering purposes in accordance with the Unified Soil Classification System (USCS). Soils sharing a common USCS classification possess similar engineering properties, including strength, permeability, and compressibility, so the USCS is useful for specifying soil types to achieve a desired performance. When quantified data regarding *LL*, *PI*, and gradation are available, the soil can be classified using ASTM D2487. However, it is prudent when recovering soil specimens from a test boring to classify the soil in the field as it is logged. In this section, a method is presented to allow engineers and technicians to quickly perform USCS soil classification using a qualitative "visual-manual" approach. These field-derived soil classifications can later be confirmed as needed in the office using results from laboratory testing.

7.3. DEFINITIONS AND THEORY

The USCS allows soil to be classified based on its engineering properties, including strength, permeability, and compressibility. To use the USCS, information regarding the *LL*, *PI*, and gradation of the soil is required. Using the USCS, each soil is assigned a two-letter group symbol and a group name. The three basic soil types and the group symbols that fall under each soil type are:

- *Gravels: GP, GW, GM, and GC,*
- *Sands: SP, SW, SM, and SC, and*
- *Silts and Clays: ML, CL, CH, MH, OH, and OL.*

There is no direct distinction between silts and clays using the USCS, although clay particles are smaller than silt particles and are mineralogically different than silt particles. Silts and clays are indirectly distinguished in the USCS through the use of *LL* and *PI* as described later. Although there are six group symbols listed under silts and clays, the last three symbols (MH, OH, and OL) are relatively uncommon.

Each group symbol has two letters. The first letter describes the soil type as follows:

- *G = gravel,*
- *S = sand,*
- *M = silt (muck),*

- *C = clay, and*
- *O = organic.*

The second letter is a modifier that provides additional description of the soil:

- *P = poorly graded,*
- *W = well graded,*
- *M = silty,*
- *C = clayey,*
- *L = low-plasticity (lean), and*
- *H = high-plasticity (fat).*

In addition to the group symbol, each soil is assigned a group name, which further modifies and describes the soil.

7.4.   EQUIPMENT AND MATERIALS

USCS field classification is qualitative in nature, but there are several items that facilitate field classification, including:

- Squeeze bottle with tap water;
- Metric ruler;
- Small bowl with mixing knife;
- Hand lens; and
- Plastic limit materials (frosted glass and 1/8-in. rod).

7.5.   PROCEDURE

7.5.1. Coarse-Grained Soils

Unlike the laboratory method described in ASTM D2487, the visual-manual procedure described in ASTM D2488 for classifying soil is qualitative and, as a result, somewhat subjective. However, soil classification using ASTM D2488 does rely on the uncorrected blow count derived from the standard penetration test (SPT) as described in ASTM D1586. Coarse-grained soil is named using the following format:

USCS group name (USCS group symbol), color, moisture, consistency, particle size, modifiers

Each portion of this name is described in the following paragraphs.

USCS group name and USCS group symbol. The USCS group name and group symbol are based on the estimated grain size, gradation, and amount of minor constituents. For coarse-grained soils, there are a total of 16 different group symbols and 32 different group names as described in Table 7.1. Selection of a group symbol and group name is largely qualitative.

Table 7.1—USCS group symbols and group names for coarse-grained soils.

Group Symbol	Group Name	
GRAVELS	**<15% sand**	**> 15% sand**
GW	Well-graded gravel	Well-graded gravel with sand
GP	Poorly-graded gravel	Poorly-graded gravel with sand
GW-GM	Well-graded gravel with silt	Well-graded gravel with silt and sand
GW-GC	Well-graded gravel with clay	Well-graded gravel with clay and sand
GP-GM	Poorly-graded gravel with silt	Poorly-graded gravel with silt and sand
GP-GC	Poorly-graded gravel with clay	Poorly-graded gravel with clay and sand
GM	Silty gravel	Silty gravel with sand
GC	Clayey gravel	Clayey gravel with sand
SANDS	**<15% gravel**	**>15% gravel**
SW	Well-graded sand	Well-graded sand with gravel
SP	Poorly-graded sand	Poorly-graded sand with gravel
SW-SM	Well-graded sand with silt	Well-graded sand with silt and gravel
SW-SC	Well-graded sand with clay	Well-graded sand with clay and gravel
SP-SM	Poorly-graded sand with silt	Poorly-graded sand with silt and gravel
SP-SC	Poorly-graded sand with clay	Poorly-graded sand with clay and gravel
SM	Silty sand	Silty sand with gravel
SC	Clayey sand	Clayey sand with gravel

Color. Use your judgment. Many engineers use a Munsell color chart to classify soil. The Munsell chart is a small blue book with pages and pages of different coded colors similar to the type of book you might look at when picking a color to paint your house. A more practical alternative may be to pick basic colors or variations of colors that are descriptive and universal (e.g. dark gray, reddish brown, etc.).

Moisture. Assessment of the moisture content in the soil is qualitative, but the three possible categories are fairly distinct from one another. Moisture in the soil is largely a function of whether the specimen was recovered from above or below the water table. Moisture categories are described in Table 7.2.

Table 7.2—Moisture description for coarse-grained soils.

Description	Criteria
Dry	Absence of moisture, dusty, dry to the touch
Moist	Damp but no visible water
Wet	Visible free water, usually soil is below the water table

Consistency. The consistency is based on the uncorrected SPT blow count. As a soil boring is logged, the drillers are likely performing SPT testing and sampling using a split-spoon sampler at regular (e.g. 5-ft) intervals, so SPT data will be available for incorporation into the field boring log. The relationship between blow count and consistency of coarse-grained soils is described in Table 7.3.

Table 7.3—Consistency for coarse-grained soils.

Blow Count	0-4	4-10	10-30	30-50	>50
Consistency	Very Loose	Loose	Med. Dense	Dense	Very Dense

Grain Size. Grain size is based on the size or range of sizes of the average particle in the soil. It is subjective, but use of a ruler or other type of scale is helpful. Grain size categories are described in Table 7.4.

Table 7.4—Grain size categories for classification of coarse-grained soils.

Category	Sieve Size	Grain Size
Fine Sand	#200 - #40	0.075-0.425 mm
Medium Sand	#40 - #10	0.425-2.0 mm
Coarse Sand	#10 - #4	2.0-4.75 mm
Fine Gravel	#4 - ¾ in.	4.75-19 mm
Coarse Gravel	¾ in. – 3 in.	19-75 mm

Modifiers. Modifiers are optional and are used to describe the quantity of a minor component in a soil, such as the amount of silt in a sand, or the amount of sand in a gravel. Modifying words are given in Table 7.5 based on percentage.

Table 7.5—Modifying terms based on percentage of constituent.

Percentage	<5	5-10	15-25	30-45	50-100
Modifier	Trace	Few	Little	Some	Mostly

Examples: The following are examples of names assigned to soils using ASTM D2487. These names would be written under the "Description" column of a soil boring log.

- *Poorly Graded Sand (SP), gray, wet, very dense, medium sand, trace silt;*

- *Poorly Graded Sand with Silt (SP-SM), gray with orange streaks, wet, dense, fine sand, few silt, trace coarse sand, petrochemical odor; or*

- *Silty Sand (SM), olive, moist, dense, fine sand, some silt.*

7.5.2. Fine-Grained Soil

Classification of fine-grained soils is mostly based on the estimate of the plasticity of the soil. Since liquid and plastic limit testing cannot be easily performed in the field, estimation of soil plasticity is based on several diagnostic tests. Fine-grained soil is named using the following format:

USCS group name (USCS group symbol), color, moisture, consistency, plasticity, modifiers

Each portion of this name is described in the following paragraphs.

USCS group name and USCS group symbol. The USCS group name and group symbol are based on the soil plasticity and amount of minor constituents. For the most common types of fine-grained soils (lean clays, silts, and fat clays), there are a total of 3 different group symbols and 21 different group names as described in Table 7.6. Selection of the USCS symbol is based on the results from five field tests. Selection of the USCS name is qualitative based on the estimated content and composition of coarse-grained material.

Deciding whether the soil is a CL, ML, or CH is based on the results from the Dry Strength, Dilatency, and Toughness Tests, which are described in ASTM D2487. The Hand Wash and Mudcrack Tests, developed by the Author, are also helpful. The final decision on plasticity is based on a preponderance of evidence. In general, different fine-grained soil types will demonstrate the responses described in Table 7.7. Each of the five diagnostic tests is described in the following paragraphs.

Dry Strength. This test involves assessing the strength of a ball of dry soil. Higher plasticity will result in a stronger soil ball when you try to break it. Criteria for the dry strength test are given in Table 7.8.

Dilatency. This test involves assessing the dilative properties of some wet soil in your hand when you shear it. Place some wet soil in your hand so that it has a "livery" appearance with a sheen of surface water. When you cup your hand, you shear the soil. If it dilates, the pore volume will increase, and the sheen will disappear as the water seeps into the soil to fill the voids. Lower-plasticity soils will be more dilative. Criteria for the Dilatency test are given in Table 7.9.

Toughness. This test involves the assessment of how tough a small ball of soil is when it is at or near the plastic limit (*PL*). Higher-plasticity soils are tougher to knead when they are at the plastic limit. Criteria for the toughness test are given in Table 7.10.

Hand Wash. This test involves smearing soil on your hands and washing it off. Higher-plasticity soils are harder to get off your hands. Criteria for the Hand Wash Test are given in Table 7.11.

Mud Crack. This test involves the assessment of how much the mudballs shrunk when they dried. Higher-plasticity soils are more susceptible to shrinking during drying, and will have more and wider mud cracks. Criteria for the Mud Crack Test are given in Table 7.12.

Table 7.6—USCS group symbols and group names for fine-grained soils.

USCS Symbol	USCS Name			
Silts				
ML	>70% fines	>85% fines		Silt
		75-85% fines	%sand>%gravel	Silt with sand
			%sand<%gravel	Silt with gravel
	<70% fines	%sand>%gravel	<15% gravel	Sandy silt
			>15% gravel	Sandy silt with gravel
		%sand<%gravel	<15% sand	Gravelly silt
			>15% sand	Gravelly silt with sand
Lean Clays				
CL	>70% fines	>85% fines		Lean clay
		75-85% fines	%sand>%gravel	Lean clay with sand
			%sand<%gravel	Lean clay with gravel
	<70% fines	%sand>%gravel	<15% gravel	Sandy lean clay
			>15% gravel	Sandy lean clay with gravel
		%sand<%gravel	<15% sand	Gravelly lean clay
			>15% sand	Gravelly lean clay with sand
Fat Clays				
CH	>70% fines	>85% fines		Fat clay
		75-85% fines	%sand>%gravel	Fat clay with sand
			%sand<%gravel	Fat clay with gravel
	<70% fines	%sand>%gravel	<15% gravel	Sandy fat clay
			>15% gravel	Sandy fat clay with gravel
		%sand<%gravel	<15% sand	Gravelly fat clay
			>15% sand	Gravelly fat clay with sand

Field Classification of Soil

Table 7.7—Responses of fine-grained soil types to diagnostic tests.

USCS Symbol	Dry Strength	Dilatency	Toughness	Hand Wash	Mudcrack
ML	None-low	Slow-Rapid	Low	Rinse	None
CL	Medium-High	None-Slow	Medium	Rub	Small
CH	High-Very High	None	High	Scrub	Large

Table 7.8—Dry strength test criteria for fine-grained soils.

Strength	Observation
None	Ball crumbles into powder when handled
Low	Ball crumbles into powder with some finger pressure
Medium	Ball breaks into pieces with considerable finger pressure
High	Must place ball between thumb and hard surface to break
Very High	Will not break between thumb and hard surface

Table 7.9—Dilatency test for fine-grained soils.

Dilatency	Observation
None	No reaction
Slow	Water disappears slowly
Rapid	Water disappears rapidly

Table 7.10—Toughness test for fine-grained soils.

Toughness	Observation
Low	Thread and ball are weak and soft
Medium	Thread and ball have some stiffness
High	Thread and ball have considerable stiffness

Table 7.11—Hand wash test for fine-grained soils.

Method	Observation
Rinse	Soil comes off under running water
Rub	Soil comes off under running water with some rubbing
Scrub	Running water and rubbing will leave some soil on your hands. You will need to scrub.

Table 7.12—Mudcrack test for fine-grained soils.

Cracks	Observation
None	No mudcracks
Few	A few little hairline cracks
Many	Cracks with widths of a millimeter or more

**Color.** Use your judgment. Many engineers use a Munsell color chart to classify soil. The Munsell chart is a small blue book with pages and pages of different coded colors similar to the type of book you might look at when picking a color to paint your house. A simpler alternative is to pick basic colors or variations of colors that are descriptive and universal (e.g. dark gray, reddish brown, etc.).

**Moisture.** Assessment of the moisture content in the soil is identical for fine-grained and coarse-grained soils. It is qualitative, but the three possible categories are fairly distinct from one another (Table 7.13). Moisture in the soil is largely a function of whether the specimen was recovered from above or below the water table. Moisture categories are described in the following table.

Table 7.13—Moisture description for fine-grained soils.

Description	Criteria
Dry	Absence of moisture, dusty, dry to the touch
Moist	Damp but no visible water
Wet	Visible free water, usually soil is below the water table

**Consistency.** The consistency is based on the uncorrected SPT blow count. As a soil boring is logged, drillers typically perform SPT testing and sampling using a split-spoon sampler at regular (e.g. 5-ft) intervals, so SPT data will be available for incorporation into the field boring log. Consistency can also be based on how easy it is to penetrate a thumbnail into an undisturbed soil specimen. The relationship between blow count and consistency of coarse-grained soils is described in Table 7.14.

Table 7.14—Consistency description for fine-grained soils.

Consistency	Qualitative Description	SPT Blow Count
Very Soft	Thumb will penetrate soil more than 1 in.	0-2
Soft	Thumb will penetrate soil approx. 1 in.	2-4
Firm	Thumb will indent soil approx. ¼ in.	4-15
Hard	Thumb will not indent soil but readily indented by thumbnail	15-30
Very Hard	Thumbnail will not indent soil	>30

**Plasticity.** Soil plasticity is based on results from the plastic limit test as described in Table 7.15.

**Modifiers.** Modifiers are optional and are used to describe the quantity of a minor component in a soil, such as the amount of sand in a fat clay. Modifying words are given in Table 7.15.

**Examples:** The following are examples of names assigned to soils using ASTM D2487. These names would be written under the "Description" column of a soil boring log form.

- *Lean Clay (CL), blue-gray, wet, firm, medium plasticity, trace fine sand;*

- *Silt with Sand (ML), olive gray, moist, very hard, nonplastic, little fine sand, trace shell fragments*

Table 7.15—Plasticity description for fine-grained soils.

Description	Criteria	Typical Soil Types
Nonplastic	A 1/8 in. thread cannot be rolled at any water content	ML
Low	The thread can barely be rolled and the lump cannot be formed when drier than the plastic limit	CL
Medium	The thread is easy to roll and not much time is required to reach the plastic limit. The lump crumbles when drier than the plastic limit	CL
High	It takes considerable time rolling and kneading to reach the plastic limit. The thread can be rerolled several times after reaching the plastic limit. The lump can be formed without crumbling when drier than the plastic limit	CH

Table 7.16—Modifying terms based on percentage of constituent.

Percentage	<5	5-10	15-25	30-45	50-100
Modifier	Trace	Few	Little	Some	Mostly

7.6. LIKELY SOURCES OF ERROR

Since soil classification using the visual manual procedure is subjective, there may be a range in soil classifications for a given soil type depending on the experience and the tendencies of the individual performing the classification. After soil specimens have been recovered in the field and returned to the laboratory with the field boring logs, it is prudent to perform an internal peer-review of the soil classifications within your company for completeness and accuracy.

7.7. ADDITIONAL CONSIDERATIONS

As mentioned previously, soils sharing a common USCS group symbol possess similar engineering properties. Table 7.17 summarizes soil types that provide various performance.

For fine-grained soils, it may not be practical to perform the Dry Strength and Mudcrack tests due to time constrains. In the time it takes to dry the mudballs and

Field Classification of Soil

perform the test, it may be possible to take some of the soil to a field laboratory and perform liquid limit and plastic limit testing to assess the soil plasticity. In addition, results of the Dilatency Test can be difficult to interpret. The Toughness and Hand Wash tests are, in the opinion of the Author, generally the best tests for assessing fine-grained soils in the field.

Table 7.17—USCS soil types and soil performance.

To achieve:	Use	Feature
Low permeability	ML, CL, CH	Fine-grained
High permeability	GP, SP	Poorly-graded
High strength	GW, SW	Well-graded
Low compressibility	GM, GP, GW	Gravelly

7.8.   SUGGESTED EXERCISES

1) Complete the Geotechnical Field Boring Log at the end of the chapter (additional data sheets can be found on the CD-ROM that accompanies this manual).to classify the soils provided by the instructor using the ASTM D2488 field classification system. This work was done as part of the Boston Central Artery Project at the South Boston Train Station. A 4.5-in. diameter borehole was installed at a surface elevation of 10 feet above mean sea level (M.S.L.) using the rotary wash drilling method on September 24, 2002. You performed the work as a subcontractor to MEK Engineering, and hired ABC Drilling to drill and log the borehole.

GEOTECHNICAL FIELD BORING LOG

Client:	Project:
Boring no:	Boring location:
Surface elevation:	Drill equipment:
Drill method:	Borehole diameter:
Total depth:	Drilled by:
Date started:	Date completed:
Logged by:	Checked by:

Depth	Sampler Type	Blows/6 in.	SPT N Blows/ft	Description	Sample Number	Recovery (in./in.)	Date	Time

8. LABORATORY SOIL COMPACTION

8.1. APPLICABLE ASTM STANDARDS

- ASTM D698: Standard Test Method for Laboratory Compaction Characteristics of Soil Using Standard Effort (12,400 ft-lb/ft^3 (600 kN-m/m^3))

- ASTM D1557: Test Method for Laboratory Compaction Characteristics of Soil Using Standard Effort (56,000 ft-lb/ft^3 (2,700 kN-m/m^3))

8.2. PURPOSE OF MEASUREMENT

Geotechnical engineers compact fine-grained soil to improve its engineering properties. Properties such as shear strength, compressibility, and hydraulic conductivity are dependent upon the methods used to compact the soil. Compacted soil is extensively used for many geotechnical structures, including earth dams, landfill liners, highway base courses and subgrades, and embankments. To predict the performance of compacted soil, and to develop appropriate construction criteria, compaction is performed in the laboratory using standardized test methods.

8.3. DEFINITIONS AND THEORY

Soil is a porous medium consisting of soil solids and water. Dry unit weight of soil, γ_d, is defined as:

$$\gamma_d = \frac{M_s}{V} g \,,$$

(8.1)

where M_s is the mass of soil solids in a volume of soil V and g is the gravitational acceleration constant. Water content, w, is defined as:

$$w = \frac{M_w}{M_s} \,,$$

(8.2)

where M_w is the mass of the water in the soil. If a given soil type is compacted using a fixed compaction effort over a range in w, a compaction curve of γ_d versus w is derived as illustrated in Fig. 8.1. The compaction curve is a concave-downward curve. At low w, γ_d is relatively low because the soil particles are in a poorly organized, flocculated configuration. The clay particles adhere to one another because they have negative charges on their faces and positive charges on their edges. As water is added and w increases, the water neutralizes some of the charge and allows the particles to disperse and assume a more organized orientation. At some point, γ_d reaches a maximum. The water content at this point is referred to as the optimum water content, w_{opt}. If more water is added, the water molecules fill in between the clay particles, and γ_d decreases. If the compaction effort increases, the compaction curve shifts up and to the left. The maximum value for γ_d increases, and w_{opt} decreases.

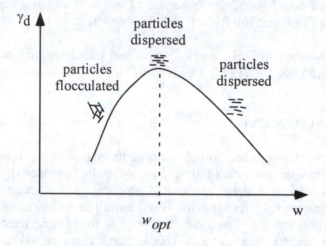

Fig. 8.1—Compaction curve showing clay particle orientations along different portions on the curve.

Dry unit weight can also be expressed as a function of w as follows:

$$\gamma_d = \frac{G_s \gamma_w}{1 + (w G_s / S)},$$
(8.3)

where G_s and γ_w are the specific gravity of soil solids and unit weight of water, respectively, and S is the degree of saturation. If S is 100%, this relationship reduces to:

$$\gamma_d = \frac{G_s \gamma_w}{1 + w G_s}.$$
(8.4)

When the relationship between γ_d and w at S=100% is superimposed onto the compaction curves, it is referred to as a Zero Air Voids (ZAV) Curve. The ZAV Curve is an excellent way to assess the validity of compaction data. In the γ_d-w domain, lines of constant S can be drawn as shown in Fig. 8.2. The portion of the compaction curve wet of w_{opt} is roughly parallel the ZAV Curve, but offset slightly. Points on the compaction curve that wet of w_{opt} generally possess S in the range of 90-93%.

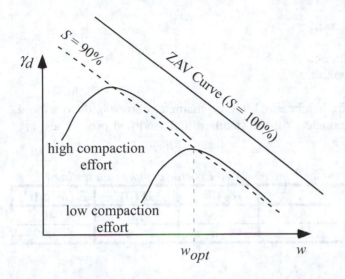

Fig. 8.2—Relative position of compaction curves and ZAV curve for low and high compaction effort (optimum water content shown for low compaction effort).

8.4.   EQUIPMENT

The following equipment is required for laboratory compaction testing:

- Soil passing the #4 sieve (5 lbs per compacted specimen);
- Standard proctor compaction hammer;
- Modified proctor compaction hammer;
- 4.0-in. diameter compaction mold, collar, and stand;
- large screwdriver;
- spray lubricant;
- ruler;
- cutting bar;
- scale capable of measuring to the nearest 0.01 g;
- drying oven;
- 3 soil moisture containers; and
- permanent marker.

Equipment used to perform laboratory compaction is illustrated in Fig. 8.3. To perform laboratory compaction, standard or modified proctor compaction effort is used. Proctor compaction consists of dropping a known weight from a fixed height onto the soil for a fixed number of drops. The total amount of energy imparted to the soil, E, is expressed as:

$$E = \frac{BLW_h D}{V_m},$$

(8.5)

where:

$B =$ number of drops per lift;

$L =$ number of lifts (layers);
$W_h =$ hammer weight;
$D =$ hammer drop height; and
$V_m =$ compaction mold volume.

The standard mold size of 4.584 in. height and 4.000 in. diameter corresponds to a mold volume, V_m, of $^1/_{30}$ ft^3. Other parameters for the standard and modified proctor test are summarized in Table 8.1.

Table 8.1—*Parameters for standard and modified Proctor compaction testing.*

Test Method	B	L	W_h (lbs)	D (ft)	E (ft-lb/ft^3)
Standard	25	3	5.5	1.0	12,400
Modified	25	5	10	1.5	56,000

Fig. 8.3—*Left to right:*
  a) standard proctor hammer;
  b) compaction mold; and
  c) modified proctor hammer.

8.5.   PROCEDURE[1]

To develop an entire compaction curve, at least five soil specimens should be compacted, with at least two specimens dry of w_{opt} and two specimens wet of w_{opt}. The following procedure, which describes how to compact one specimen, would be repeated for each specimen.

---

[1] Don't forget to visit www.wiley.com/college/kalinski to view the lab demo!

1)   Obtain a 5 - lb sample of moist soil that has been prepared by the instructor.

2)   Assemble the compaction mold, stand, and collar. Measure and record the height and diameter of the mold. Tighten the wingnuts and seat all the pieces together properly, and spray the inside of the assembly with an aerosol spray lubricant.

3)   Compact the soil in lifts (3 for standard, 5 for modified). Use 25 blows/lift and scarify the compacted surface between lifts with the screwdriver. Place enough loose soil into the mold prior to compaction of each lift such that the compacted material will occupy approximately one-third or one-fifth of the mold (depending on the compaction effort). The top of the final lift should be just above the top of the mold such that it will need to be trimmed slightly.

4)   Remove the collar and trim the excess soil off the top of the mold (Fig. 8.4).

Fig. 8.4—Trimming excess soil from the top of the mold using the cutting bar.

5)   Extrude the specimen (Fig. 8.5) and obtain the net mass of the compacted soil in pounds (Note: 1.00 lb = 454 grams).

6)   Obtain samples from the top, middle, and bottom of the specimen, and perform water content measurements on the samples to obtain the average water content, w_{avg}, for the specimen.

7)   Calculate the dry unit weight of the soil, γ_d, in pcf:

$$\gamma_d = \frac{Mg}{(1 + w_{avg})V_m},$$ (8.5)

where M is in pounds and g is equal to 32.2 ft/s^2.

Fig. 8.5—Extruding the compacted specimen from the mold using an extruder

8.6. EXPECTED RESULTS

For most soils, optimum w_{opt} is typically around 14-18% and 10-15% for standard and modified proctor compaction effort, respectively. Maximum dry unit weight is typically around 100-110 pcf and 120-130 pcf for standard and modified proctor compaction effort, respectively. The degree of saturation of soil wet of w_{opt} is typically around 90-93%.

8.7. LIKELY SOURCES OF ERROR

In ASTM D698 and D1557, it is recommended that when dry soil is combined with water to achieve a target moisture content, the mixture should be allowed to sit and hydrate for least 16 hours prior to compaction. Complete hydration is important, and inadequate hydration time may result in erroneous results and a poorly defined compaction curve. For this exercise, your instructor should prepare the soil specimens in advance and provide you with moist soil at the desired target moisture contents for testing.

When compacting several specimens to fully define a compaction curve, it is also important to use new soil for each specimen. If a compacted specimen is crushed, re-hydrated, and recompacted for a second specimen, γ_d will be slightly elevated.

8.8. ADDITIONAL CONSIDERATIONS

The engineering properties of soil are dependent upon whether soil is compacted wet or dry of w_{opt}. The strength of soil is much higher when dry of w_{opt}, while the hydraulic conductivity of soil is much lower when wet of w_{opt}. Therefore, if high strength or low hydraulic conductivity is desired, an acceptance window of γ_d-w pairs can be established as shown in Fig. 8.6.

ASTM D698 and D1557 each describe 3 procedures for compacting soil (Table 8.2). Herein, Procedure A is described. Other than the parameters listed in Table 8.2, the three procedures are identical. The choice of procedure is based on the amount of coarse material present in the soil that is retained by the #4, $^3/_8$ in., or ¾ in. sieves as follows:

- *Procedure A, B, or C:*   May be used if $P_{+\#4} < 20\%$;
- *Procedure B or C:*   May be used if $P_{+\#4} > 20\%$ and $P_{+3/8 \text{ in.}} < 20\%$;
- *Procedure C:*   May be used if $P_{+3/8 \text{ in.}} > 20\%$ and $P_{+3/4 \text{ in.}} < 30\%$.

Procedure C has a larger compaction volume, but requires more blows per lift. As a net result, each procedure imparts the same amount of energy to the specimen.

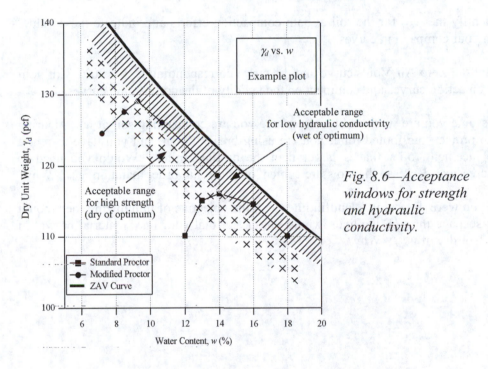

Fig. 8.6—*Acceptance windows for strength and hydraulic conductivity.*

Table 8.2—*Description of procedures for performing compaction testing.*

Procedure	Mold Dia. (in.)	Mold Vol. (ft^3)	Fraction of material to use	Blows per lift
A	4.00	$^1/_{30}$	passing #4	25
B	4.00	$^1/_{30}$	passing $^3/_8$ in.	25
C	6.00	$^1/_{13.33}$	passing $^3/_4$ in.	56

8.9.   SUGGESTED EXERCISES

1)   Perform one standard and one modified proctor compaction test on the soil provided to you by the instructor using the Compaction Test Data Sheets at the end of the chapter (additional data sheets can be found on the CD-ROM that accompanies this manual). Report your results to the instructor in a timely manner.

2)   Plot compaction curves for both the standard and modified proctor compaction tests using all the data points from the class provided to you by the instructor. Use the Compaction Curve Plot Data Sheet at the end of the chapter (additional data sheets can be found on the CD-ROM that accompanies this manual).

3)   Identify the w_{opt} for the soil at both compaction efforts and indicate these values on your compaction curves.

4)   Plot the Zero Air Voids curve and the curve corresponding to $S = 90\%$ with your compaction curves and comment on the validity of the compaction curves.

5)   Assume you are working at a site where you are compacting soil at w_{opt}. During the project, the footed roller you are using breaks down, and you have to rent a replacement roller that is larger than the original roller. Would you need to increase or decrease the moisture content of the soil in order maintain w_{opt}? Why?

6)   If you were designing a landfill liner with the objective of minimizing the amount of seepage through the liner, would you prefer that your liner material be wet of w_{opt} or dry of w_{opt}? Why?

COMPACTION TEST (ASTM D698, D1557)
LABORATORY DATA SHEET

I. GENERAL INFORMATION

Tested by:	Date tested:
Lab partners/organization:	
Client:	Project:
Boring no.:	Recovery depth:
Recovery date:	Recovery method:
Soil description:	

II. TEST DETAILS

Compaction effort (standard or modified):	
Soil hydration period prior to compaction:	Max. particle size:
Compaction procedure (A, B, or C):	Mold diameter:
Mold height:	Mold volume (V_m):
Notes, observations, and deviations from ASTM D698 and D1557 test standards:	

III. MEASUREMENTS AND CALCULATIONS

Location Within Specimen	Top	Middle	Bottom
Container ID			
Mass of container (M_c)			
Mass of moist soil + container (M_1)			
Mass of dry soil + container (M_2)			
Moisture Content (w)			
Average Water Content (w_{avg})			

Net Mass of Compacted Specimen (M):	Dry Unit Weight (γ_d):

IV. EQUATIONS AND CALCULATION SPACE

$$w = \frac{M_1 - M_2}{M_2 - M_c} \, x\, 100\%$$

$$\gamma_d = \frac{Mg}{(1 + w_{avg})V_m}$$

COMPACTION TEST (ASTM D698, D1557)
LABORATORY DATA SHEET

I. GENERAL INFORMATION

Tested by:	Date tested:
Lab partners/organization:	
Client:	Project:
Boring no.:	Recovery depth:
Recovery date:	Recovery method:
Soil description:	

II. TEST DETAILS

Compaction effort (standard or modified):	
Soil hydration period prior to compaction:	Max. particle size:
Compaction procedure (A, B, or C):	Mold diameter:
Mold height:	Mold volume (V_m):
Notes, observations, and deviations from ASTM D698 and D1557 test standards:	

III. MEASUREMENTS AND CALCULATIONS

Location Within Specimen	Top	Middle	Bottom
Container ID			
Mass of container (M_c)			
Mass of moist soil + container (M_1)			
Mass of dry soil + container (M_2)			
Moisture Content (w)			
Average Water Content (w_{avg})			

Net Mass of Compacted Specimen (M):	Dry Unit Weight (γ_d):

IV. EQUATIONS AND CALCULATION SPACE

$$w = \frac{M_1 - M_2}{M_2 - M_c} \, x \, 100\%$$

$$\gamma_d = \frac{Mg}{(1 + w_{avg})V_m}$$

COMPACTION CURVE PLOT (ASTM D698, D1557)

I. GENERAL INFORMATION

Tested by:	Date tested:
Lab partners/organization:	
Client:	Project:
Boring no.:	Recovery depth:
Recovery date:	Recovery method:
Soil description:	

II. TEST DETAILS

Compaction effort (standard or modified):	
Compaction procedure (A, B, or C):	Specific Gravity of Soil Solids (G_s):
Notes, observations, and deviations from ASTM D698 and D1557 test standards:	

III. MEASUREMENTS AND CALCULATIONS

Standard Proctor (ASTM D698)		Modified Proctor (ASTM D1557)		ZAV Curve	
w	γ_d	w	γ_d	w	γ_d

IV. EQUATION AND CALCULATION SPACE

ZAV: $\gamma_d = \dfrac{G_s \gamma_w}{1 + w G_s}$

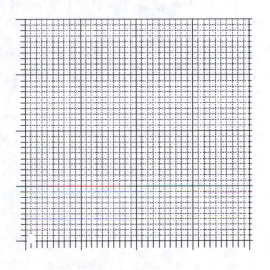

Dry Unit Weight, γ_d (   )

Moisture Content, w (   )

9. FIELD MEASUREMENT OF DRY UNIT WEIGHT AND MOISTURE CONTENT

9.1. APPLICABLE ASTM STANDARDS

- ASTM D1556: Standard Test Method for Density and Unit Weight of Soil in Place by the Sand-Cone Method

- ASTM D2167: Standard Test Method for Density and Unit Weight of Soil in Place by Rubber Balloon Method

9.2. PURPOSE OF MEASUREMENT

The sand-cone and rubber balloon tests are destructive in situ field tests used to measure the total unit weight (γ) of compacted earth materials. When accompanied with moisture content (w) measurements of the same material, the sand cone and rubber balloon tests can be used to measure both w and dry unit weight (γ_d) to confirm that the earth materials are compacted in accordance with construction specifications.

9.3. DEFINITIONS AND THEORY

9.3.1. Overview

When soil is used to construct highway subgrades and base courses, waste containment liners, earth dams, embankments, and other purposes, the soil must be compacted in accordance with construction specifications. Specifications for compacted soil are typically given in terms of an acceptable range of moisture content (w) and/or dry unit weight (γ_d) based on results of laboratory compaction tests (ASTM D698 and D1557).

To confirm that soil is compacted in accordance with construction specifications, γ and w of representative samples of compacted soil are measured as part of a Construction Quality Assurance (CQA) plan. A CQA plan specifies the type and frequency of laboratory or field tests to be performed on the soil, as well as acceptance criteria. Moisture content and dry unit weight are often specified. Given γ and w, γ_d is expressed as:

$$\gamma_d = \frac{\gamma}{1+w}, \tag{9.1}$$

where w is expressed as a decimal.

To measure γ and w in situ, a small hole (on the order of 0.1 ft^3) is excavated at the surface of a compacted layer of soil. The soil is removed, and its moisture content is measured using a standard method such as ASTM D2216. The mass of the soil, M_{wet}, is also recorded. Equation 9.1 can be rewritten in terms of w and M_{wet}:

$$\gamma_d = \frac{(M_{wet})g}{V(1+w)},\qquad\qquad(9.2)$$

where g is the gravitational constant (i.e. 9.81 m/s^2), and V is the volume of the hole. Therefore, any method that provides a means to measure V will be useful in deriving γ_d for CQA purposes when accompanied with moisture content measurements of the material removed from the hole. There are two such methods commonly used today: the sand cone method (ASTM D1556) and the rubber balloon method (ASTM D2167). Each method is described in the following sections.

9.3.2. Sand Cone Method

The sand cone method employs the use of poorly graded sand that, when poured out of a container through funnel into a hole, fills the hole at a known, pre-calibrated value for γ_d. By weighing the container before and after the hole is filled, the volume of the hole can be calculated based on the calibrated value for γ_d.

The sand cone device is illustrated in Fig. 9.1. The device consists of a sand container, funnel, and sand. The sand must be a clean, dry, poorly graded sand with a coefficient of uniformity ($C_u = D_{60}/D_{10}$) less than 2.0, a maximum particle size (D_{100}) less than 2.0 mm, and less than 3% by weight passing the #60 (250 μm) sieve. The sand should consist of rounded or subrounded particles rather than angular particles. The sand should be stored in an airtight container between tests so that it remains dry.

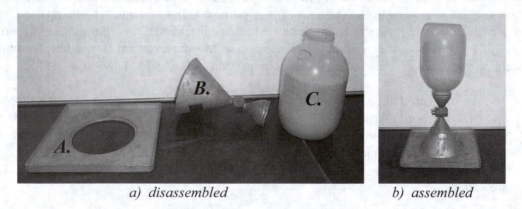

a) disassembled   b) assembled

Fig. 9.1—Sand cone device. Parts include A) base plate, B) funnel, and C) sand container.

9.3.3. Rubber Balloon Test

The rubber balloon test employs the use of a water-filled graduated cylindrical chamber (Fig. 9.2). The chamber sits on a metal bottom piece with a hole in the center. A rubber balloon is fixed to the hole, and the entire configuration is sealed and airtight. The water in the chamber can be placed under pressure or vacuum by pumping a reversible bulb by hand. To perform a test, the device is placed on a base plate over an excavated test hole, and the water is placed under pressure. The water-filled balloon is pushed out of the cylinder to fill the test hole. The volume of the test hole is taken as the change in water level in the cylinder.

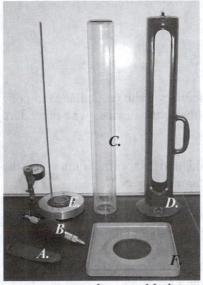

Parts include:
A) rubber balloon
B) reversible hand pump
C) graduated cylinder
D) device housing
E) metal bottom piece
F) base plate

a. disassembled   *b. assembled*

9.4.   EQUIPMENT AND MATERIALS

9.4.1. Sand Cone Test

The following equipment and materials are required for performing the sand cone test:

- Small digging tools (e.g. shovels, trowels, chisels, etc.);
- large sealable plastic bag or airtight container;
- poorly graded subrounded to rounded sand;
- sand cone device, including container and funnel;
- scale capable of measuring to the nearest 1.0 g; and
- base plate.

9.4.2. Rubber Balloon Test

The following equipment and materials are required for performing the rubber balloon test:

- Small digging tools (e.g. shovels, trowels, chisels, etc.);

- assembled rubber balloon device;
- reversible bulb hand pump;
- scale capable of measuring to the nearest 1.0 g; and
- base plate.

9.5.   PROCEDURE[1]

9.5.1. Sand Cone Test

9.5.1.1. Calibration of the Sand Cone Device

Since the results of sand cone testing are highly dependent upon the particular sand cone device and type of sand used, it is very important to calibrate the device. The procedure for calibrating the device is as follows:

1) Fill the sand cone container with dry sand and place the funnel on the container. Record the mass of the filled sand cone device, M_6.

2) Place the base plate on a clean, flat surface and place the inverted sand cone device over the base plate (Fig. 9.1b). Open the valve in the funnel and allow the sand to fill the base plate and funnel. Close the valve after the base plate and funnel are filled. Remove the sand cone device from the base plate and record the mass of the device with the remaining sand, M_7.

3) Calculate the mass of the sand in the base plate and funnel, M_2:

$$M_2 = M_6 - M_7.$$ (9.1)

4) Refill the container and obtain the mass of the refilled device (M_8). Place the base plate over a calibration container of known volume. Many base plates are machined to snugly fit over a proctor mold with a known volume, V_1, of either $^1/_{13.33}$ or $^1/_{30}$ ft^3, so a proctor mold may be used to facilitate the calibration.

5) Place the inverted sand cone device over the base plate, open the valve, and fill the base plate, funnel, and calibration chamber with sand (Fig. 9.3). After the calibration chamber, base plate, and funnel are filled, close the valve. Remove the sand cone device from the base plate and weigh the sand cone device with the remaining sand, M_9.

6) Calculate the mass of the sand in the calibration chamber, M_5:

$$M_5 = M_8 - M_9 - M_2.$$ (9.2)

---

[1] Don't forget to visit www.wiley.com/college/kalinski to view the lab demo!

Fig. 9.3 – Filling the base plate, funnel, and calibration chamber with sand for calibration.

7) Calculate the total unit weight of the sand, γ_1:

$$\gamma_1 = \frac{M_5 g}{V_1},$$ (9.3)

where g is the gravitational constant.

9.5.1.2. Performing a Sand Cone Measurement

Once the sand and sand cone device have been calibrated using the procedure described in Section 9.5.1.1., sand cone measurements can be performed using the following procedure:

1) Fill the sand cone device with the same type of sand used for the calibration. Obtain the mass of the filled sand cone, M_{10}.

2) Locate a flat, level spot on the surface of the material to be tested. Place the base plate on the surface.

3) Excavate a test hole through the center of the base plate (Fig. 9.4). The minimum test hole volume is dependent upon the maximum particle size as described in Table 9.1. The shape of the test hole should approximate the

shape of the calibration chamber. The base plate should not overhang the test hole, and the bottom of the test hole should be flat or concave upward. Place the excavated soil in a sealed plastic bag to use for measurement of moisture content later.

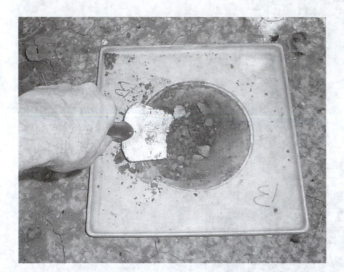

Fig. 9.4 – Excavation of a test hole through the center of the base plate.

Table 9.1 – Minimum test hole volume based on maximum particle size.

Maximum Particle Size (in.)	Minimum Test Hole Volume (ft³)
0.5	0.050
1.0	0.075
1.5	0.100

4) Position the filled sand cone device over the excavated test hole. Open the valve and fill the test hole, base plate, and funnel with sand. Do not perform the test if there are significant ambient vibrations (e.g. heavy equipment operation), and take care not to move or shake the device during filling. After filling, close the valve and measure the mass of the sand cone with the remaining sand, M_{11}.

5) Calculate the mass of the sand used to fill the test hole, funnel, and base plate, M_1:

$$M_1 = M_{10} - M_{11}. \qquad (9.4)$$

6) Calculate volume of the test hole, V:

$$V = \frac{(M_1 - M_2)g}{\gamma_1}. \qquad (9.5)$$

7) Record the moist mass of the material excavated from the test hole, M_3.

8) Dry the soil in an oven using the methods described in ASTM D2216 to obtain the dry mass of the soil, M_4. Calculate the moisture content of the material, w:

$$w = \frac{M_3 - M_4}{M_4} x100\% . \tag{9.6}$$

9) Calculate the dry unit weight, γ_d, of the soil:

$$\gamma_d = \frac{M_4 g}{V} . \tag{9.7}$$

9.5.2. Rubber Balloon Test

9.5.2.1. Calibration of the Rubber Balloon Device

Like the sand cone device, it is also important to calibrate the rubber balloon device to obtain accurate, consistent measurements. The procedure for calibration of the rubber balloon device prior to testing is as follows:

1) Assemble the rubber balloon device by filling the cylinder with water, fixing the rubber membrane inside the bottom piece, and fixing the bottom piece to the cylinder (Fig. 9.2b).

2) Set the base plate on a flat surface and position the filled device over the base plate. Pressurize the water in the rubber balloon device using the hand bulb pump until the water level in the cylinder reaches a constant level and the balloon has completely filled the base plate (Fig. 9.5). Record the water level, V_o, in the cylinder. Apply sufficient reaction load to the device (e.g. have an assistant hold the device down) so that the pressure of the balloon does not lift the device up off the base plate.

3) Reverse the flow direction of the bulb pump, and apply vacuum to the water until the membrane is pulled back into the bottom of the device.

9.5.2.2. Performing a Rubber Balloon Measurement

Once the rubber balloon device has been calibrated using the procedure described in Section 9.5.2.1., rubber balloon measurements can be performed using the following procedure:

1) Locate a flat, level spot on the surface of the material to be tested. Place the base plate on the surface, and fix the base plate to the surface using nails or spikes.

Fig. 9.5 – Calibrating the rubber balloon device on a flat surface.

2) Excavate a test hole through the center of the base plate. The minimum test hole volume is dependent upon the maximum particle size as described in Table 9.1. The base plate should not overhang the test hole, and the bottom of the test hole should be flat or concave upward. There should be no sharp edges in the test hole that may puncture the balloon. The excavated soil should be sealed in a plastic bag to use for measurement of moisture content later.

3) Position the rubber balloon device over the base plate. Pressurize the water in the rubber balloon device using the hand bulb pump until the water level in the cylinder reaches a constant level and the balloon has completely filled the base plate. Record the water level, V, in the cylinder. Apply sufficient reaction load to the device (e.g. have an assistant hold the device down) so that the pressure of the balloon does not lift the device up off the base plate.

4) Calculate the volume of the test hole, V_h:

$$V_h = V - V_o. \tag{9.8}$$

5) Obtain the mass of the moist soil excavated from the test hole, M_{wet}. Dry the soil in an oven using the methods described in ASTM D2216 to obtain the dry mass of the soil, M_d. Calculate the moisture content of the soil, w:

$$w = \frac{M_{wet} - M_d}{M_d} \, x \, 100\% \, . \tag{9.9}$$

6) Calculate the dry unit weight, γ_d, of the soil:

$$\gamma_d = \frac{M_d g}{V_h} \, . \tag{9.10}$$

9.6. EXPECTED RESULTS

Dry unit weight can range from 100 to 130 pcf for compacted soils. For projects that involve soil compaction, specifications typically state that soil should be compacted to within 90% of maximum dry unit weight for standard or modified proctor compaction effort. Maximum dry unit weight is typically around 100-110 pcf and 120-130 pcf for standard and modified proctor compaction effort, respectively.

9.7. LIKELY SOURCES OF ERROR

For the sand cone test, error may occur if vibrations are present during the calibration or measurement process because vibrations tend to densify sand. For the rubber balloon test, error may occur if the device leaks, or if the balloon is punctured due to angular particles protruding into the test hole. For either method, error may occur if the base plate hangs over the edge of the test hole, or if the bottom of the test hole is not flat or concave upward. In these instances, the test hole may not be completely filled during the measurement.

9.8. ADDITIONAL CONSIDERATIONS

The sand cone and rubber balloon methods are not suitable for soils that are susceptible to deformation, including organic, saturated, or highly plastic soils. The methods are also not suitable for cohesionless soils because test holes excavated in cohesionless material will not remain open.

The sand cone and rubber balloon methods are typically specified as part of CQA programs for earthwork projects. These tests are destructive tests, which may not be desirable for structures such as compacted waste containment liners, where field performance is dependent on the integrity of the liner. As a result, these tests may be prescribed at a relatively infrequent interval (e.g. one test for every 20,000 cubic yards of compacted material), and supplemented with a nondestructive test at a more frequent interval (e.g. one test for every 1,000 cubic yards of compacted material). The most common nondestructive method for measuring γ_d and w in situ is the nuclear gauge

(ASTM D2922, D3017) as illustrated in Fig. 9.6. A radioactive neutron source (^{241}Am) irradiates the soil with neutrons, which are detected with a neutron detector. Since hydrogen atoms in water molecules absorb neutrons, w is proportional to the amount of neutrons detected. A separate gamma radiation source (^{137}Cs) irradiates the soil with gamma radiation, which is detected with a gamma detector. Denser materials scatter gamma radiation (Compton scattering), so the amount of gamma radiation detected is related to the unit weight of the soil. Calibration curves are used to correlate the levels of neutrons and gamma radiation detected by the nuclear gauge to γ_d and w.

The nuclear gauge is widely used, but there is significant administrative effort and burden associated with managing the nuclear material. As a result, recent effort has been dedicated to the development of nondestructive devices that do not rely on the use of nuclear material. One such method applies time domain reflectometry (ASTM D6780), which uses measurements of the electrical conductive and capacitive properties of the soil to estimate γ_d and w.

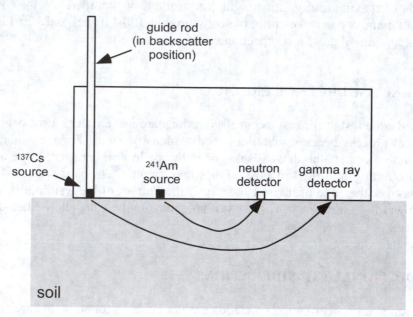

Fig. 9.6 – Nuclear gauge for in situ estimate of γ_d and w.

9.9. SUGGESTED EXERCISES

1) Assemble and calibrate a sand cone device, and perform a sand cone measurement on a pad of compacted soil. Use the Measurement of In Situ Dry Unit Weight and Moisture Content Using the Sand Cone Method data sheet at the end of the chapter (additional data sheets can be found on the CD-ROM that accompanies this manual).

2) Assemble and calibrate a rubber balloon device, and perform a rubber balloon measurement on the same pad of compacted soil. Use the Measurement of In Situ Dry Unit Weight and Moisture Content Using the Rubber Balloon Method data sheet at the end of the chapter (additional data sheets can be found on the CD-ROM that accompanies this manual).

3) Compare the results between the two methods, and discuss any similarities, differences, or likely sources of error.

SAND CONE TEST (ASTM D1556)
FIELD DATA SHEET

I. GENERAL INFORMATION

Tested by:	Date tested:
Lab partners/organization:	
Client:	Project:
Field compaction method:	Date material compacted:
Soil description:	

II. TEST DETAILS

Description of sand used in sand cone (particle shape, C_u, D_{100}, %-#60):
Description of calibration chamber (shape and dimensions):

Calibration chamber volume (V_1):	Max. particle size of compacted material:

Notes, observations, and deviations from ASTM D1556 test standard:

III. MEASUREMENTS AND CALCULATIONS

Calibration	Measurement
Mass of filled device (M_6):	Mass of filled device (M_{10}):
Mass of device after filling base plate and funnel (M_7):	Mass of device after filling base plate, funnel, and test hole (M_{11}):
Mass of sand in the base plate and funnel (M_2):	Mass of sand in the base plate, funnel, and test hole (M_1):
Mass of refilled device (M_8):	Volume of test hole (V):
Mass of refilled device after filling base plate, funnel, and calibration chamber (M_9):	Mass of moist material excavated from the test hole (M_3):
Mass of sand in the calibration chamber (M_5):	Dry mass of material excavated From the test hole (M_4):
Total unit weight of the sand (γ_1):	

Moisture content (w):	Dry unit weight (γ_d):

IV. EQUATIONS AND CALCULATION SPACE

$$M_2 = M_6 - M_7 \qquad M_1 = M_{10} - M_{11} \qquad w = \frac{M_3 - M_4}{M_4} \times 100\%$$

$$M_5 = M_8 - M_9 - M_2 \qquad V = \frac{(M_1 - M_2)g}{\gamma_1} \qquad \gamma_d = \frac{M_4 g}{V}$$

$$\gamma_1 = \frac{M_5 g}{V_1}$$

RUBBER BALLOON TEST (ASTM D2167)
FIELD DATA SHEET

I. GENERAL INFORMATION

Tested by:	Date tested:
Lab partners/organization:	
Client:	Project:
Field compaction method:	Date material compacted:
Soil description:	

II. TEST DETAILS

Description of rubber balloon device (manufacturer, serial no.):	
Approximate operating pressure:	Max. particle size of compacted material:
Notes, observations, and deviations from ASTM D2167 test standard:	

III. MEASUREMENTS AND CALCULATIONS

Water level in cylinder without test hole (V_o):
Water level in cylinder with test hole (V):
Volume of test hole (V_h):
Mass of moist material excavated from the test hole (M_{wet}):
Dry mass of material excavated from the test hole (M_d):
Moisture content (w):
Dry unit weight (γ_d):

IV. EQUATION AND CALCULATION SPACE

$$V_h = V - V_o$$

$$w = \frac{M_{wet} - M_d}{M_d} \, x \, 100\%$$

$$\gamma_d = \frac{M_d g}{V}$$

10. MEASUREMENT OF HYDRAULIC CONDUCTIVITY OF GRANULAR SOILS USING A FIXED WALL PERMEAMETER

10.1. APPLICABLE ASTM STANDARDS

- ASTM D2434: Standard Test Method for Permeability of Granular Soils (Constant Head)

10.2. PURPOSE OF MEASUREMENT

It is important to quantify the volume of groundwater flow from areas of high potential to low potential. This information is useful in estimating the performance of landfill liners, the migration of contaminated groundwater, and other applications. To quantify flow through soil, the hydraulic conductivity (a.k.a. permeability) of the soil must be known. Hydraulic conductivity of granular soil, including sands and gravels, is measured in the laboratory using a fixed-wall permeameter. In this exercise, hydraulic conductivity will be used using the constant head and falling head test methods.

10.3. DEFINITIONS AND THEORY

10.3.1. Darcy's Law and the Constant Head Test

Water moves through soil in accordance with Darcy's Law. Given a cylinder of soil with length L and cross-sectional area A subjected to a constant head difference of Δh (Fig. 10.1), the rate of flow through the soil, q, can be expressed as:

$$q = k \frac{\Delta h}{L} A .$$ (10.1)

In this expression, k is the hydraulic conductivity of the soil. The flow rate q can be expressed as flow volume Q per unit time t,

$$q = \frac{Q}{t} .$$ (10.2)

The ratio of Δh to L is defined as the hydraulic gradient i:

$$i = \frac{\Delta h}{L} ,$$ (10.3)

such that Darcy's Law can be rewritten as:

$$q = kiA. \tag{10.4}$$

Darcian velocity, v_D, can be expressed as:

$$v_D = ki, \tag{10.5}$$

and Darcy's Law can also be expressed as:

$$q = v_D A. \tag{10.6}$$

Darcian velocity is also referred to as discharge velocity. Darcian velocity is not equal to seepage velocity, v_s. Darcian velocity is always less than v_s, and the two terms are related by porosity, n:

$$v_s = \frac{v_D}{n}. \tag{10.7}$$

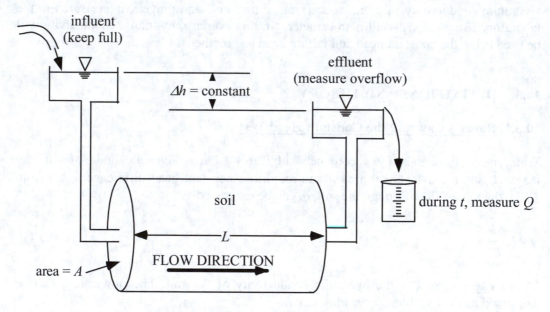

Fig. 10.1—Flow of water through soil under constant head conditions.

Darcy's Law is based on the assumption of laminar flow. Under conditions of laminar flow, k is independent of i. To ensure laminar flow, ASTM D2434 states that i should be in the range of 0.2-0.3 and 0.3-0.5 for loose soils and dense soils, respectively. The low end of these ranges corresponds to coarser soils, while the high end of these ranges corresponds to finer soils. This assumption can be validated by performing the test over a range of i, and creating a plot of q versus iA. The slope of this curve is k. For lower values of i, flow is laminar, the relationship between q and iA is linear, and k is independent of i. For higher values of i, flow becomes turbulent and the relationship between q and iA becomes nonlinear. When applying laboratory test results, k should be measured under a hydraulic gradient representative of anticipated field conditions regardless of flow regime to obtain appropriate results.

10.3.2. Falling Head Test

The constant head test is described in ASTM D2434. The falling head test is not included as part of the ASTM D2434 standard, but is a simpler test because 1) it does not require a water source to keep the influent reservoir at a constant level and 2) measurement of flow volume is not necessary. The test configuration (Fig. 10.2) includes a fixed-wall permeameter and a supply reservoir with a cross-sectional area a. By measuring the head at the beginning of the test, H_1, and at the end of the test, H_2, after a permeation period of t, k can be calculated.

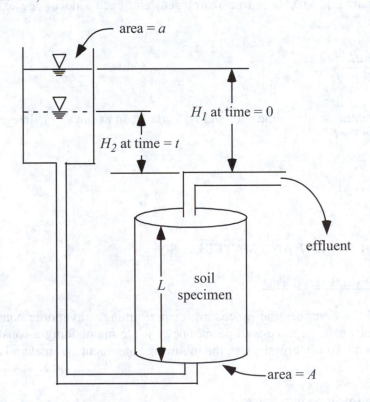

Fig. 10.2 – Flow of water through soil under falling head conditions.

Darcy's Law is used to derive the expression for k using the falling head test method. For the falling head test configuration, Darcy's Law can be expressed as:

$$q = k\frac{H}{L}A,$$

(10.8)

where H is the instantaneous head across the specimen. Instantaneous flow rate, q, can be expressed as a function of the incremental change in head Δh during an increment in time Δt:

$$q = a\frac{\Delta h}{\Delta t} .$$

(10.9)

Combining Eqns. 10.8 and 10.9:

$$a\frac{dh}{dt} = k\frac{H}{L}A .$$

(10.10)

The terms in Eqn. 10.10 can be rearranged, and each side of the expression can be integrated:

$$\int_0^t k\,dt = \int_{H_2}^{H_1}\left(\frac{aL}{A}\frac{dh}{H}\right),$$

(10.11)

where H_1 is the head at time $= 0$, and H_2 is the head at time $= t$. Integrating both sides and solving for k,

$$k = \frac{aL}{At}\ln\left(\frac{H_1}{H_2}\right).$$

(10.12)

10.4. EQUIPMENT AND MATERIALS

10.4.1. Constant Head Test

ASTM D2434 describes the procedure for performing a hydraulic conductivity test of granular soils using a fixed-wall permeameter while maintaining a constant head across the specimen. To perform the test, the following equipment and materials are required:

- Coarse-grained soil;
- fixed-wall permeameter;
- constant-elevation water reservoir;
- tap water source;
- 2 manometers;
- measuring tape or yardstick;
- large vessel for collecting effluent;
- scale capable of measuring to the nearest 1.0 g;
- timing device capable of measuring to the nearest second; and
- vacuum source capable of achieving a vacuum of 500 mm Hg (-9.67 psi).

The fixed-wall permeameter is illustrated in Fig. 10.3, and the overall constant head test configuration is illustrated in Fig. 10.4. All four ports on the permeameter should have valves that can be closed. To prevent dislodging of soil particles at the top of the specimen, a spring is placed between the top cap and the top porous stone or wire screen. The spring should apply 5-10 pounds of force to the specimen. The spacing between manometer ports, L_c, should be greater than the specimen diameter D. The purpose of the porous stones or wire screens is to prevent particle migration during permeation, but it is also helpful to place filter paper between the soil and the porous stones or wire screens. If porous stones are used, the hydraulic conductivity of the porous stones should be greater than the hydraulic conductivity of the soil.

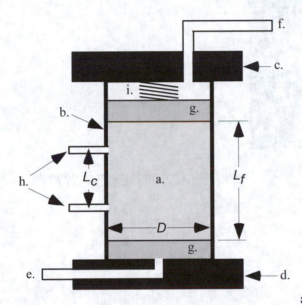

a. cylindrical soil specimen
b. permeameter side wall
c. permeameter top cap
d. permeameter bottom cap
e. influent port
f. effluent port
g. porous stone or wire screen
h. manometer ports
i. spring w/ 5 to 10 lb force

Fig. 10.3—Illustration of the fixed-wall permeameter (schematic and photograph).

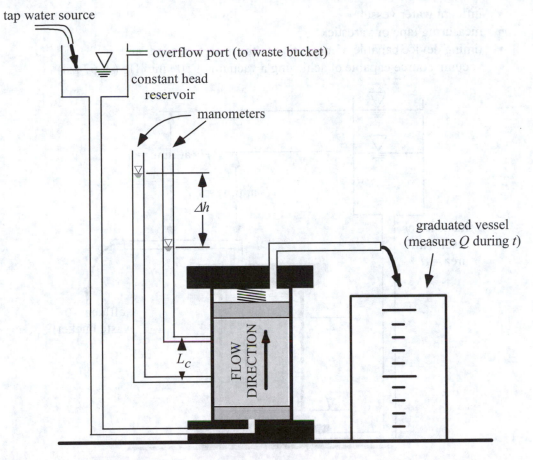

Fig. 10.4—Constant-head hydraulic conductivity test configuration.

For all soils, the fraction retained by the ¾ in. sieve should be removed prior to testing. The minimum required diameter of the specimen, D, is dependent on the maximum particle size of the fraction passing the ¾ in. sieve, and the percent retained by the #10 sieve (2.00 mm) or the $^3/_8$ in. sieve, as detailed below:

- If max. particle size is bet. 2.00 mm and $^3/_8$ in. and $P_{+\#10} < 35\%$ → $D > 3.0$ in.
- If max. particle size is bet. 2.00 mm and $^3/_8$ in. and $P_{+\#10} > 35\%$ → $D > 4.5$ in.
- If max. particle size is bet. $^3/_8$ in. and ¾ in. and $P_{+3/8\ in.} < 35\%$ → $D > 6.0$ in.
- If max. particle size is bet. $^3/_8$ in. and ¾ in. and $P_{+3/8\ in.} > 35\%$ → $D > 9.0$ in.

10.4.2. **Falling Head Test**

The falling head test (Fig. 10.5) is a simpler alternative to the constant head. To perform a falling head test, the following equipment and materials are required.

- Coarse-grained soil;
- fixed-wall permeameter;

- influent water vessel;
- measuring tape or yardstick;
- timing device capable of measuring to the nearest second; and
- vacuum source capable of achieving a vacuum of 500 mm Hg (-9.67 psi).

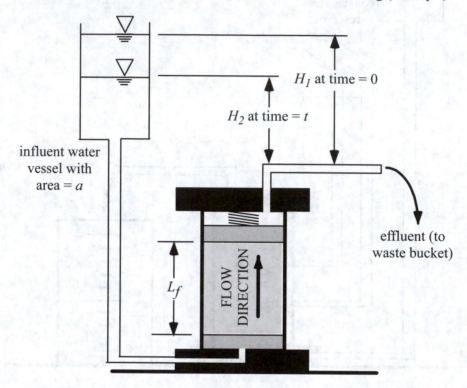

Fig. 10.5—Falling-head hydraulic conductivity test configuration.

The same permeameter can be used for both the constant head and falling head tests, but the manometers are not used for the falling head test and the manometer valves should remain closed. It is also important to note that the length term is different for the constant and falling head tests. For the constant head test, the length L_c is the distance between the manometer ports. For the falling head test, the length L_f is the total length of the soil specimen.

10.5. PROCEDURE[1]

10.5.1. Constant Head Test

1) Obtain a soil-filled permeameter from your instructor and assemble the constant-head test configuration as shown in Fig. 10.4. Measure the distance between the

---

[1] Don't forget to visit www.wiley.com/college/kalinski to view the lab demo!

manometer ports (L_c) and the diameter of the soil specimen in the permeameter (D). Calculate the specimen cross-sectional area A:

$$A = \frac{\pi D^2}{4}. \qquad (10.13)$$

2) Soil must be saturated for Darcy's Law to be valid. Saturate the specimen using the following procedure:
   a. Close the influent valve and manometer valves.
   b. Apply a 500-mm Hg (9.67-psi) vacuum to the effluent port for 15 min.
   c. Open the influent valve and allow water to saturate the specimen.
   d. Close the influent and effluent valves and remove the vacuum.

3) Open the influent valve, effluent valve, and manometer valves, and begin permeating tap water through the specimen while maintaining a constant head. Once the water level in the manometers has stabilized, record effluent flow volume Q during time t. If your vessel is not graduated, you can calculate Q using the conversion factor 1.00 cm^3 = 1.00 g. For coarse-grained soil, you should be able to permeate a sufficient amount of water (a few liters) in about 10 minutes or so. Also measure the corresponding head loss Δh between the two manometers.

4) Repeat Step 3 a total of 4 times. For each trial, vary i so that the tests span a range in i from 0.2 to 0.5.

5) Calculate k for each trial using the following relationship:

$$k = \frac{Q L_c}{\Delta h A t}. \qquad (10.14)$$

6) Create a plot q versus iA using the four test points to identify the laminar and turbulent flow regimes.

10.5.2. Falling Head Test

1) Assemble the falling head test configuration as shown in Fig. 10.5 using the permeameter from the constant head test. Measure the length of the specimen (L_f) and the diameter of the specimen in the permeameter (D). Calculate the specimen cross-sectional area A. Close the two manometer valves and leave them closed for the experiment.

2) Soil must be saturated for Darcy's Law to be valid. Saturate the specimen using the following procedure:
   a. Close the influent valve and manometer valves.
   b. Apply a 500-mm Hg (9.67-psi) vacuum to the effluent port for 15 min.
   c. Open the influent valve and allow water to saturate the specimen.

    d.   Close the influent and effluent valves and remove the vacuum.

3) Calculate the cross-sectional area of the falling head water reservoir, a.

4) Measure the initial head, H_1.

5) Open the valves and permeate water through the specimen. Record the time, t, required for the head to drop to H_2.

6) Repeat Steps 4 and 5 a total of 4 times. For each trial, vary the initial hydraulic gradient $i_i = H_1/L_f$ so that the tests span a range in i_i from 0.2 to 0.5.

7) Calculate the hydraulic conductivity of the specimen, k, in cm/s for each trial using the following relationship:

$$k = \frac{aL_f}{At} ln\left(\frac{H_1}{H_2}\right). \tag{10.15}$$

10.6. EXPECTED RESULTS

The fixed-wall permeameter tests described in this exercise are intended for coarse-grained granular soils with less than 10% fines, which include the USCS group symbols SP, SW, GP, GW, SP-SM, SP-SC, GP-GM, and GP-GC. For these soils, k is typically on the order of 10^{-2} to 10^{-3} cm/s. Coarse-grained soils with greater than 10% fines, including SM, SC, GM, and GC, typically have k on the order of 10^{-5} to 10^{-6} cm/s. Fine-grained soils, including ML, CL, and CH, typically have k on the order of 10^{-6} to 10^{-8} cm/s.

10.7. LIKELY SOURCES OF ERROR

While minor measurement errors may occur in a laboratory-scale measurement of hydraulic conductivity due to errors in measurement of length, volume, weight, or time, the most likely error is in how laboratory measurements are applied to field conditions. Laboratory-measured k may be several orders of magnitude less than actual large-scale field values. Specimens in the laboratory are not always representative of field conditions, and macroscopic features that govern field permeation such as gravel lenses, root holes, and animal burrows, may not be present in laboratory specimens. As a result, it may be necessary to measure field-scale hydraulic conductivity in situ using field methods such as the sealed double-ring infiltrometer (ASTM D5093).

10.8. ADDITIONAL CONSIDERATIONS

The tests performed in this exercise are intended for soils possessing less than 10% fines. For soils containing greater than 10% fines, the flexible-wall permeability test (ASTM

D5084) is used. The flexible-wall permeameter consists of a cylindrical soil specimen placed inside a latex membrane, which is placed inside a chamber of pressurized water. If a fine-grained soil specimen is placed in a fixed-wall permeameter, the diameter of the specimen cannot conform to the inside diameter of the permeameter, and water may seep along the perimeter of the specimen without permeating through the soil. This behavior is referred to as sidewall leakage, which leads to erroneously high values for k. The pressure in the chamber creates intimate contact between the membrane and the soil specimen to minimize sidewall leakage.

Since the soil specimen in a flexible-wall permeameter is under pressure, the influent and effluent water can also be placed under pressure. Pressurization of the influent greatly increases the hydraulic gradient and flow rate in the specimen, which is necessary to obtain measurable amounts of effluent in a reasonable amount of time due to the low hydraulic conductivity of fine-grained soils. Pressurization of the influent and effluent increases the pore pressure in the specimen, which shrinks air bubbles in the specimen and increases the degree of saturation. This practice is referred to as backpressure saturation.

For this exercise, it is assumed that the instructor has prepared the soil specimens in the fixed-wall permeameter in advance. However, ASTM D2434 provides detailed guidance for sample preparation methods. For practical applications, the samples should be prepared to be consistent with design criteria. For example, if gravel is to be used as a drainage layer for a leachate collection and removal system in a landfill, it may be specified to be placed at a given void ratio or dry unit weight. The gravel sample to be tested in the laboratory should therefore be prepared at the same void ratio or dry unit weight.

10.9. SUGGESTED EXERCISES

1) Conduct four constant head hydraulic conductivity tests and report your results on the Hydraulic Conductivity of Granular Soil Under Constant Head Data Sheet at the end of the chapter (additional data sheets can be found on the CD-ROM that accompanies this manual). Vary the amount of head as described in Section 10.6.1 so that measurements are performed over a range in i from 0.1 to 0.5. Report your answer in scientific notation with two significant figures in cm/s (e.g. $k = 1.4 \times 10^{-3}$ cm/s).

2) Conduct four falling head hydraulic conductivity tests and report your results on the Hydraulic Conductivity of Granular Soils Under Falling Head Data Sheet at the end of the chapter (additional data sheets can be found on the CD-ROM that accompanies this manual). Vary the amount of head as described in Section 10.6.2 so that measurements are performed over a range in i from 0.1 to 0.5. Report your answer in scientific notation with two significant figures in cm/s (e.g. $k = 1.4 \times 10^{-3}$ cm/s).

3)   Using the four constant head test points, create a plot q versus iA and identify the laminar and turbulent flow regimes. Comment on the range in i over which the flow is laminar and Darcy's Law is valid.

4)   Compare the values for k derived using the constant and falling head tests, and comment on the similarities or differences in the results.

HYDRAULIC CONDUCTIVITY OF GRANULAR SOIL
UNDER CONSTANT HEAD (ASTM D2434)
LABORATORY DATA SHEET

I. GENERAL INFORMATION

Tested by:	Date tested:
Lab partners/organization:	
Client:	Project:
Boring no.:	Recovery depth:
Recovery date:	Recovery method:
Soil description:	

II. TEST DETAILS

Max. particle size:	$P_{+\#10}$ or $P_{+3/8\ in}$ (state which):
Specimen diameter, D:	Specimen area, A:
Manometer port spacing, L_c:	Specimen length:
Dry mass of soil, M_s:	Volume of soil, V:
Specific gravity of soil solids, G_s:	Dry unit weight, γ_d:
Void ratio, e:	Scale type/serial no./precision:
Saturation vacuum level:	Saturation vacuum duration:
Specimen preparation method:	
Notes, observations, and deviations from ASTM D2434 test standard:	

III. MEASUREMENTS AND CALCULATIONS

Test No.	Head Loss (Δh)	Hydraulic Gradient (i)	Flow Volume (Q)	Time (t)	Flow Rate (q)	Hydraulic Conductivity (k)

IV. EQUATION AND CALCULATION SPACE

$$A = \frac{\pi D^2}{4} \qquad q = \frac{Q}{t}$$

$$i = \frac{\Delta h}{L_c} \qquad k = \frac{Q L_c}{\Delta h A t}$$

HYDRAULIC CONDUCTIVITY OF GRANULAR SOIL
UNDER FALLING HEAD
LABORATORY DATA SHEET

I. GENERAL INFORMATION

Tested by:	Date tested:
Lab partners/organization:	
Client:	Project:
Boring no.:	Recovery depth:
Recovery date:	Recovery method:
Soil description:	

II. TEST DETAILS

Max. particle size:	$P_{+\#10}$ or $P_{+3/8\,in}$ (state which):
Specimen diameter, D:	Specimen area, A:
Influent reservoir area, a:	Specimen length, L_f:
Dry mass of soil, M_s:	Volume of soil, V:
Specific gravity of soil solids, G_s:	Dry unit weight, γ_d:
Void ratio, e:	Scale type/serial no./precision:
Saturation vacuum level:	Saturation vacuum duration:
Specimen preparation method:	
Notes and observations:	

III. MEASUREMENTS AND CALCULATIONS

Test No.	Initial Head (H_1)	Initial Hydraulic Gradient (i_i)	Final Head (H_2)	Time (t)	Hydraulic Conductivity (k)

IV. EQUATION AND CALCULATION SPACE

$$A = \frac{\pi D^2}{4} \qquad i_i = \frac{H_1}{L_f}$$

$$k = \frac{aL_f}{At} ln\left(\frac{H_1}{H_2}\right)$$

11. ONE-DIMENSIONAL CONSOLIDATION TEST OF COHESIVE SOIL

11.1. APPLICABLE ASTM STANDARDS

- ASTM D2435: Standard Test Method for One-Dimensional Consolidation Properties of Soils

11.2. PURPOSE OF MEASUREMENT

When a layer of fine-grained (cohesive) soil, including ML, CL, and CH, is subjected to an increase in effective stress through an increase overburden stress, the soil undergoes a long-term reduction in void ratio, e, which is accompanied by settlement of the soil layer. To quantify both the ultimate amount of settlement and the time rate of settlement in the soil layer, a one-dimensional consolidation test is performed in the laboratory. Using laboratory-derived parameters, field settlement behavior of the soil layer can be predicted.

11.3. DEFINITIONS AND THEORY

The vertical effective stress in a horizontal layer of fine-grained soil, σ', can be expressed as the difference between vertical total stress, σ, and the pore water pressure, u:

$$\sigma' = \sigma - u. \tag{11.1}$$

If a layer of overburden soil is placed on top of the layer of fined grained soil, σ' will increase by $\Delta\sigma$, an amount which is equal to the product of the total unit weight and thickness of the overburden layer. However, σ' does not increase instantly in the fine-grained soil layer. Initially, $\Delta\sigma$ is carried by the pore water in the soil, and excess pore water pressure is generated. Total stress and u both increase by an amount equal to $\Delta\sigma$, and the initial change in effective stress, $\Delta\sigma'$, is equal to zero. As time passes, the pressurized pore water in the fine-grained soil layer permeates into an adjacent layer of more freely-draining (i.e.; higher hydraulic conductivity) soil. As this occurs, $\Delta\sigma$ is gradually transferred from the pore water to the soil particles, and the rate at which this occurs is controlled by the hydraulic conductivity of the fine-grained soil. In addition, e decreases and the volume of the fine-grained soil layer decreases. Once the excess pore water pressure has fully dissipated, σ' has increased by an amount equal to $\Delta\sigma$. This time-dependent dissipation of excess pore water pressure and long-term reduction in e caused by the loading of fine-grained soils is referred to as consolidation.

11.4. EQUIPMENT AND MATERIALS

The following equipment and materials are required for performing a one-dimensional consolidation test:

- Undisturbed soil specimen;
- soil trimming equipment;
- petroleum jelly or vacuum grease;
- consolidation ring;
- consolidation cell;
- filter paper;
- calipers;
- consolidation load frame;
- weights;
- deformation indicator capable of measuring to the nearest 0.0001 in.
- timer;
- squeeze bottle;
- two oven-safe moisture content containers;
- oven-safe container large enough to contain the consolidation ring;
- permanent marker;
- scale capable of measuring to the nearest 0.01 g; and
- soil drying oven set at $110^{\circ} \pm 5^{\circ}$ C.

There are many different experimental configurations for performing a one-dimensional consolidation test. Some systems apply load to the soil using dead weight with a mechanical advantage, while others apply load using hydraulic or pneumatic pressure. Some systems record data automatically using computer-driven logging systems and electronic deformation indicators, while others rely on manual data recording using analog dial gauges. The equipment at your institution will probably differ from the equipment presented in the photographs herein, but the basic principles are the same. Consult with your instructor for details on to how to operate the equipment in your laboratory.

11.5. PROCEDURE[1]

11.5.1. Preparation of Soil Specimen and Configuration of Test

The one-dimensional consolidation test is performed by placing a cylindrical specimen of undisturbed fine-grained soil in a consolidation cell (Fig. 11.1). With regard to specimen dimensions, ASTM D2435 specifies that 1) the minimum height and diameter is 0.5 in. and 2.00 in., respectively, 2) the height must exceed 10 times the maximum particle size, and 3) the diameter:height ratio must exceed 2.5.

---

[1] Don't forget to visit www.wiley.com/college/kalinski to view the lab demo!

An undisturbed specimen may be recovered using a ring-lined sampler (ASTM D3550). Use of the ring-lined sampler eliminates the need to trim the specimen, and the specimen can be placed directly in the consolidation cell. However, Shelby tube sampling (ASTM D1587) is a more common method for recovering undisturbed soil specimens. Use of a Shelby tube specimen requires careful trimming of the soil to remove the outer, more disturbed portion of the soil, and to cut a soil specimen that fits into a consolidation ring. Using this approach, a consolidation ring with a beveled edge is slowly and gently pushed onto the undisturbed soil specimen with a diameter larger than the diameter of the ring. As the ring is incrementally slid down onto the soil specimen, excess soil is carefully trimmed away from the sides of the specimen so that the trimmed soil specimen fits snugly into the consolidation ring. Petroleum jelly or vacuum grease is placed inside the consolidation ring prior to trimming to reduce friction on the sides of the specimen as the soil deforms during the consolidation test. Once the ring is completely filled with soil, the soil is trimmed flush with the top and bottom of the consolidation ring, and the net weight of the soil in the consolidation ring is obtained.

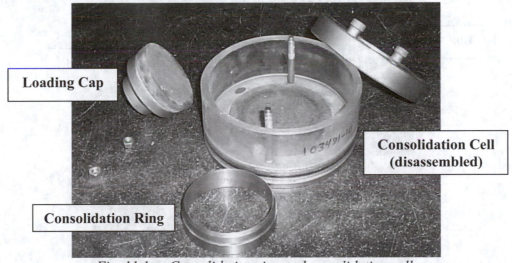

Fig. 11.1 — Consolidation ring and consolidation cell.

The soil-filled consolidation ring is then placed in the consolidation cell. The soil specimen is sandwiched between two porous stones. The bottom stone is fixed to the consolidation cell, while the top stone is fixed to the loading cap used to transfer load to the soil specimen. The porous stones act as freely draining materials so that drainage in the soil specimen is two-way and the drainage distance is half the height of the specimen. To prevent soil from intruding into the porous stones and clogging them, filter paper disks are placed between the soil and the porous stones.

The consolidation cell is filled with water and placed in the load frame (Fig.11.2). The load frame includes a loading arm and a hanger from which weights are hung (Fig. 11.3). The load frame is often configured with a mechanical advantage of around 10:1 to magnify the applied load. The weights that accompany a load frame are typically labeled with an equivalent stress that has been calculated by the manufacturer based on the mechanical advantage of the load frame and the area of the soil specimen in the consolidation cell.

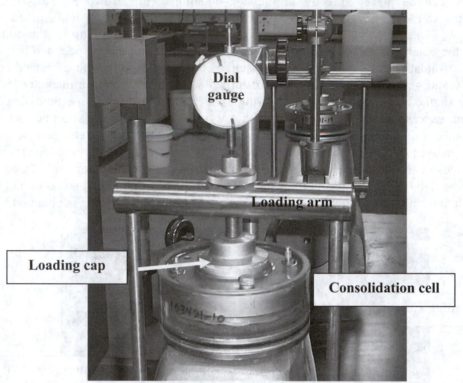

Fig. 11.2 – Consolidation cell placed in load frame.

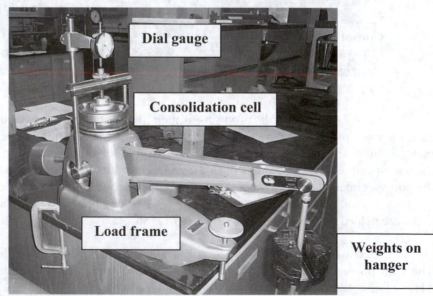

Fig. 11.3 – One-dimensional consolidation test configuration showing weights on hanger.

To record soil deformation during the test, a deformation indicator is positioned over the soil specimen. An analog dial gauge is shown in Figs. 11.2 and 11.3, but other types of devices, including digital dial gauges, proximeters, and linear variable displacement transducers (LVDTs), can also be used. Consult with your instructor for instructions on using the deformation indicators used in your laboratory.

When the consolidation cell is first placed in the load frame, a seating load of 100 psf is typically applied. Once the seating load is applied, the deformation indicator is set to zero. The seating load may be increased for high-plasticity soils where swelling is anticipated, or decreased for soft soils to minimize consolidation under the seating load.

11.5.2. Performing Laboratory Measurements

The one-dimensional consolidation test is performed by incrementally loading the soil specimen. Load increments start low and increase by a factor of two (e.g. 250 psf, 500 psf, 1000 psf, etc.). The load sequence for the entire test may also include intermediate unloading and reloading steps. By unloading and reloading the specimen, an accurate measurement of the recompression index, C_r, can be obtained. A typical loading sequence is listed in Table 11.1. In this sequence, load 7 is an unloading step, and loads 8 and 9 are reloading steps.

Table 11.1 — Typical loading sequence for a one-dimensional consolidation test.

Load Number	Applied Stress (psf)
1	250
2	500
3	1000
4	2000
5	4000
6	8000
7	2000
8	4000
9	8000
10	16000
11	32000
12	64000

The duration of each load increment may be as short as a few hours, or as long as a few days, provided that the load increment is long enough to include all primary consolidation and some secondary compression. Silts may require a shorter load increment, while clays may require a longer load increment. It is convenient, however, to use a 24-hour loading period for each load. This facilitates scheduling of the test, while more or less assuring that primary consolidation is complete before the end of each load increment. Since consolidation occurs rapidly at first and gradually slows down, readings are taken more frequently at the beginning of the load. A typical set of readings is listed in Table 11.2.

Table 11.2—Typical set of readings for a single load increment.

Reading Number	Time After Loading
1	6 sec.
2	15 sec.
3	30 sec.
4	1 min.
5	2 min.
6	4 min.
7	8 min.
8	15 min.
9	30 min.
10	1 hr.
11	2 hr.
12	4 hr.
13	8 hr.
14	24 hr.

The times listed in Table 11.2 are target times, and readings should be taken at times within about 20% of these target times. For example, it is acceptable to substitute a 35-minute reading if a reading cannot be taken exactly 30 minutes after initial loading.

There are several devices available for use on newer load frames to record deformation, including proximeters, LVDTs, and digital dial gauges. These devices can be calibrated to read out in units of inches or millimeters, and use of these devices is relatively straightforward. However, many older load frames are instrumented with analog dial gauges (Fig. 11.4). Readings are taken on an analog dial gauge in units of divisions, which are later converted to units of length using a conversion factor (e.g. 0.0001 in./division). A dial gauge consists of a large hand moving around a large face, and a smaller hand moving around a smaller inset face. The small hand advances one tick for each complete rotation of the large hand. To read the dial gauge, the large hand is read in divisions and the small hand is read in hundreds of divisions, and the two numbers are added together. When the small hand is between two numbers, the smaller of the two numbers is selected.

The deformation measured by a dial gauge or other deformation indicator includes deformation as a result of soil consolidation. However, compression of the filter paper and porous stones also contributes to the measured deformation. Machine deflection readings are made by assembling the consolidation cell with the two porous stones and two sheets of filter paper, but without the soil. The consolidation cell is placed in the load frame, filled with water, and incrementally loaded using the same loading sequence to be used in the consolidation test. Each load is placed for about one minute, and the deflection corresponding to deformation of the porous stones and filter paper is recorded. To facilitate data reduction, one deflection reading is taken for each load increment. When reducing consolidation data for a given load increment, all of the measurements are corrected for machine deflection by subtracting the corresponding machine deflection value from each measurement.

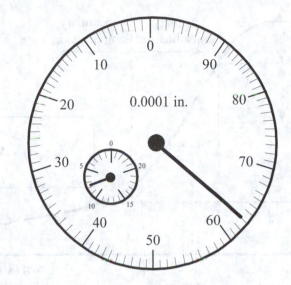

Fig. 11.4 — Dial gauge shown with a reading of 763 divisions.

11.5.3. Deriving c_v and d_{100}

11.5.3.1.   Overview

Time-deformation data are used to derive coefficients of vertical consolidation, c_v, and the deformation corresponding to the end of primary consolidation, d_{100}, for each load increment. The coefficient of vertical consolidation decreases with increasing stress as hydraulic conductivity decreases in accordance with Terzaghi's Theory of Consolidation. There are two graphical methods that can be used to derive c_v and d_{100}. The log time method is advantageous because of the availability of commercial software that can be used to create semi-log plots. The root time method is advantageous because there is less interpretation involved, and settlement is approximately proportional to the square root of time. Each method is outlined in the following sections.

11.5.3.2.   Log Time Method

The procedure for deriving c_v and d_{100} using the log time method (Fig. 11.5) is as follows:

1) Correct all deformation readings for machine deflection by subtracting the appropriate machine deflection value for the corresponding load increment.

2) Plot deformation, d, versus time, t, in minutes on a semi-log plot. Time is plotted on the log scale. If deformation is recorded using a dial gauge, d should be in

units of divisions. If deformation is recorded using a deformation indicator that reads directly in units of length, d should be in units of length.

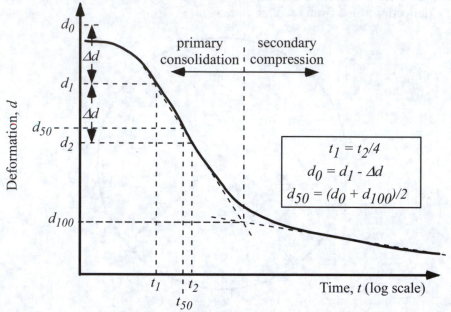

Fig. 11.5 — Time-settlement data plotted using semi-log paper using the log time method.

3) Identify d_{100} as the deformation corresponding to the intersection of the straight portions of the primary consolidation and secondary compression curves.

4) Identify t_2 as a time near the point of inflection of the primary consolidation portion of the curve.

5) Calculate $t_1 = t_2/4$.

6) Identify d_1 and d_2 corresponding to t_1 and t_2.

7) Calculate $\Delta d = d_2 - d_1$.

8) Calculate $d_0 = d_1 - \Delta d$.

9) Calculate $d_{50} = (d_0 + d_{100})/2$

10) Identify t_{50} corresponding to d_{50}.

11) Calculate the drainage distance corresponding to an average degree of consolidation of 50%, H_{D50}. If a dial gauge is used as a deformation indicator,

$$H_{D50} = \frac{H_o - d_{50}(K)}{2}, \tag{11.2}$$

where H_o is the initial height of the specimen., and K is a dial gauge conversion factor (e.g. 0.0001 in./division). If deformation is recorded using a deformation indicator that reads directly in units of length,

$$H_{D50} = \frac{H_o - d_{50}}{2}. \tag{11.3}$$

12) Calculate the coefficient of vertical consolidation, c_v:

$$c_v = \frac{T_{50}(H_{D50})^2}{t_{50}} = \frac{0.197(H_{D50})^2}{t_{50}}. \tag{11.4}$$

11.5.3.2.   Root Time Method

The procedure for deriving c_v and d_{100} using the root time method (Fig. 11.6) is as follows:

1) Correct all deformation readings for machine deflection by subtracting the appropriate machine deflection value for the corresponding load increment.

2) Plot deformation, d, versus time, t, in minutes on using root paper. Time is plotted on the root scale. If deformation is recorded using a dial gauge, d should be in units of divisions. If deformation is recorded using a deformation indicator that reads directly in units of length, d should be in units of length.

3) Identify d_0 as the y-intercept of the curve.

4) Extend linear portion of the curve to the bottom of the graph.

5) Find the intersection point of the linear portion and bottom of graph (X).

6) Multiply intersection point by 1.15 and post the 1.15X point on the bottom of graph.

7) Draw line between 1.15X point and d_0.

8) Identify d_{90} as the intersection of the line and the curve.

9) Calculate d_{100}:

$$d_{100} = d_0 + 1.11(d_{90} - d_o). \tag{11.5}$$

10) Identify t_{90} corresponding to d_{90}.

11) Calculate the drainage distance corresponding to an average degree of consolidation of 50%, H_{D50}. Use Eqn. 11.2 if a dial gauge is used as a deformation indicator. Use Eqn. 11.3 if deformation is recorded using a deformation indicator that reads directly in units of length.

12) Calculate the average degree of vertical consolidation, c_v:

$$c_v = \frac{T_{90}(H_{D50})^2}{t_{90}} = \frac{0.848(H_{D50})^2}{t_{90}}.$$

(11.6)

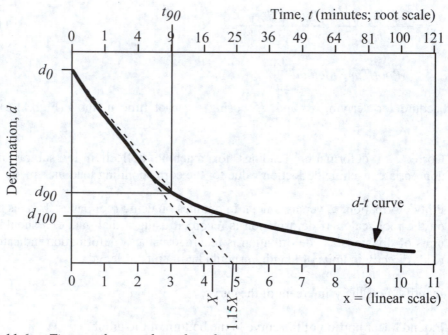

Fig. 11.6 — Time-settlement data plotted using root paper using the root time method.

11.5.4. Deriving e – log σ' Curve

The deformation at the end of primary consolidation for each load increment, d_{100}, is used to derive a curve of e versus σ'. This curve should be plotted as a semi-log plot, with e plotted on the linear axis and σ' plotted on the log axis. After the consolidation test is completed, the consolidation cell is dismantled and the dry mass of the soil, M_d, is obtained. This information is used to calculate the initial void ratio of the specimen prior to the consolidation test, e_o:

$$e_o = \frac{V_o - \dfrac{M_d}{G_s \rho_w}}{\dfrac{M_d}{G_s \rho_w}}. \tag{11.7}$$

In Eqn. 11.7, V_o is the initial volume of the specimen (i.e.; the volume of the consolidation ring) and ρ_w is the mass density of water. From this, the height of solids in the soil specimen, H_s, is obtained:

$$H_s = \frac{H_o}{1 + e_0}. \tag{11.8}$$

The d_{100} values are used to calculate the change in void ratio for each load increment at the end of primary consolidation, Δe. If a dial gauge is used to measure deformation,

$$\Delta e = \frac{\Delta H}{H_s} = \frac{d_{100}(K)}{H_s}. \tag{11.9}$$

If deformation is recorded using a deformation indicator that reads directly in units of length,

$$\Delta e = \frac{\Delta H}{H_s} = \frac{d_{100}}{H_s}. \tag{11.10}$$

Finally, the void ratio at the end of primary consolidation, e, is calculated for each load increment σ':

$$e = e_0 - \Delta e. \tag{11.11}$$

The $e - \sigma'$ data pairs are then used to plot the $e - \log \sigma'$ curve (Fig. 11.7). The $e - \log \sigma'$ curve is bilinear, with the flatter portion corresponding to reconsolidation at lower stresses, and the steeper portion corresponding to virgin consolidation at higher stresses. The $e - \log \sigma'$ curve is used to derive the compression index, C_c, the recompression index, C_r, and the maximum previous consolidation pressure, σ'_{max}, which are used to estimate ultimate settlement. The compression and recompression indices are the slopes of the two portions of the curve. To calculate these parameters, two points are selected along a linear section of each portion the curve. The two points possess void ratios e_1 and e_2, and stresses σ_1' and σ_2', respectively, and are selected so that $e_1 > e_2$ and $\sigma_2' > \sigma_1'$. Compression and recompression index are then expressed as:

$$C = \frac{e_1 - e_2}{\log \sigma_2 - \log \sigma_1}. \tag{11.12}$$

The maximum previous consolidation pressure, σ'_{max}, represents the highest vertical effective stress that the soil has ever experienced. The Casagrande method is a graphical method for deriving σ'_{max} as described in ASTM D2435, and is illustrated in Fig. 11.8.

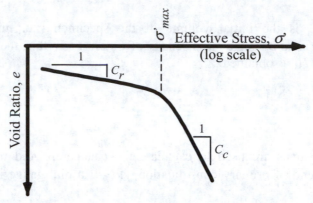

Fig. 11.7 – Typical e – log σ' curve.

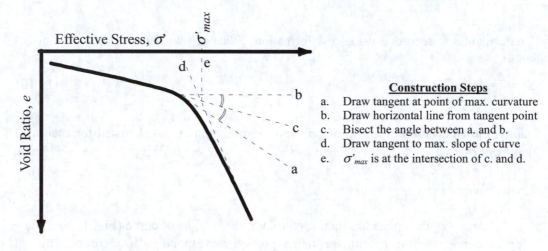

Construction Steps
a. Draw tangent at point of max. curvature
b. Draw horizontal line from tangent point
c. Bisect the angle between a. and b.
d. Draw tangent to max. slope of curve
e. σ'_{max} is at the intersection of c. and d.

Fig. 11.8 – Casagrande graphical construction method for deriving σ'_{max}.

11.6. EXPECTED RESULTS

Most soils have some degree of overconsolidation, so the $e – \log \sigma'$ curve derived from a one-dimensional consolidation test is typically bi-linear as illustrated in Fig. 11.7. Recompression index, C_r, is typically around 0.1, while compression index, C_c, may be around 0.5 to 1.0. Maximum previous consolidation pressure may be a few thousand psf, but is dependent upon the overconsolidation ratio of the soil. Overconsolidation ratio (*OCR*) is defined as:

$$OCR = \frac{\sigma'_{max}}{\sigma'_i},$$ (11.13)

Where σ'_i is the in situ effective stress of the soil. OCR ranges from 1.0 for younger, normally consolidated clays, to over 4.0 for older, highly overconsolidated clays. Coefficient of vertical consolidation, c_v, is typically around 0.1-0.5 ft^2/day for virgin consolidation (stresses lower than σ'_{max}), and around 0.5-1.0 ft^2/day for recompression (stresses greater than σ'_{max}).

11.7. LIKELY SOURCES OF ERROR

Results obtained from a one-dimensional laboratory consolidation test may be used to estimate ultimate settlement and settlement rates in the field. However, the laboratory test only allows excess pore water pressure to dissipate in the vertical direction. In the field, larger-scale heterogeneities in a soil mass may allow lateral drainage, which shortens the drainage distance and accelerates consolidation. Wick drains and sand drains are often used to accelerate consolidation in thick clay layers by shortening the drainage distance. Wick drains and sand drains are vertically oriented drainage features with diameters on the order of inches, that are installed over a grid with a spacing on the order of 10 ft.

The one-dimensional consolidation test is also based on the assumption of plane loading with no edge effects. In the field, structures such as embankments are finite in dimension, so pore water can migrate laterally as well as vertically to dissipate excess pore pressure, which also accelerates consolidation. To obtain a more accurate estimate of time rates of settlement for features with edges and finite dimensions, such as embankments or foundations, numerical methods may be employed.

Finally, soil disturbance plays a significant role in estimating ultimate settlement. Ideally, consolidation tests are performed on undisturbed soil specimens. Soil disturbance tends to reduce the distinction between the reconsolidation and virgin consolidation portions of the $e - \log \sigma'$ curve. As illustrated in Fig. 11.9, soil disturbance tends to increase C_r and reduce C_c.

11.8. ADDITIONAL CONSIDERATIONS

Consolidation occurs in soil due to long-term increases in effective stress. Increasing the overburden is one mechanism that increases effective stress by increasing total stress. However, lowering the water table also increases effective stress by reducing pore water pressure. In some areas where groundwater pumping has been extensive, large urban areas have undergone significant settlement due to an increase in effective stress. For large construction projects in congested downtown areas where the water table is near the ground surface and groundwater must be pumped to drain the basement for excavation, settlement of adjacent structures may be significant, and mitigating measures may be warranted.

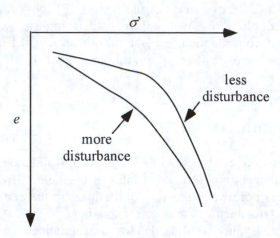

Fig. 11.9 – Effect of soil disturbance on e – log σ' curve.

11.9. EXERCISES

1) Perform a machine deflection test for each load in your test. Record the data using the Machine Deflection Test Data Sheet at the end of the chapter (additional data sheets can be found on the CD-ROM that accompanies this manual). If your instructor has already performed a machine deflection test, obtain the data from them.

2) Prepare an undisturbed specimen of fine-grained soil for performing a one-dimensional consolidation test. Record the data using the Specimen Preparation Data Sheet at the end of the chapter (additional data sheets can be found on the CD-ROM that accompanies this manual). If your instructor has already prepared the specimen for testing, obtain the data from them.

3) Record time-deformation data for one increment of loading as directed by your laboratory instructor. Record your data using the Time-Deformation Data Sheet at the end of the chapter (additional data sheets can be found on the CD-ROM that accompanies this manual).

4) Plot deformation versus time on a semi-log plot. Use the log time graphical construction method to calculate c_v and d_{100} for your load increment. Create the plot using either semi-log graph paper or commercial software, and record your data and calculations on the Time-Deformation Plotting Using the Log Time Method Data Sheet at the end of the chapter (additional data sheets can be found on the CD-ROM that accompanies this manual).

5) Plot deformation versus time using the root time plotting paper at the end of the chapter. Use the root time graphical construction method to calculate c_v and d_{100} for your load increment. Record your calculations on the Time-Deformation Plotting Using the Root Time Method Data Sheet at the end of the chapter (additional data sheets can be found on the CD-ROM that accompanies this manual).

6) Obtain d_{100} data for each load increment in the test from your instructor to create an $e - \log \sigma'$ plot. Create the plot using either semi-log graph paper or commercial software, and record your data and calculations on the Construction of $e - \log \sigma'$ Curve Data Sheet at the end of the chapter (additional data sheets can be found on the CD-ROM that accompanies this manual). Calculate C_c, C_r, and σ'_{max}.

ONE-DIMENSIONAL CONSOLIDATION TEST (ASTM D2435)
MACHINE DEFLECTION MEASUREMENTS
LABORATORY DATA SHEET

I. GENERAL INFORMATION

Test performed by:	Date tested:
Lab partners/organization:	
Load frame type/serial no.:	
Load duration:	Blank material and thickness:
Filter paper type:	
Porous stone type and thickness:	
Deformation indicator type and conversion factor K (if applicable):	
Notes, observations, and deviations from ASTM D2435 test standard:	

II. MEASUREMENTS

Pressure (psf)	Deformation Reading ( )

ONE-DIMENSIONAL CONSOLIDATION TEST (ASTM D2435)
SPECIMEN PREPARATION MEASUREMENTS
LABORATORY DATA SHEET

I. GENERAL INFORMATION

Specimen prepared by:	Date:
Lab partners/organization:	
Client:	Project:
Boring no.:	Recovery depth:
Recovery date:	Recovery method:
Soil description:	

II. TEST DETAILS

Load frame type/serial no.:	
Scale type/serial no./precision:	
Consolidation ring diameter:	Initial specimen height, H_o:
Consolidation ring mass:	Specimen volume, V_o:
Specific gravity of soil solids, G_s:	
Notes, observations, and deviations from ASTM D2435 test standard:	

III. MEASUREMENTS AND CALCULATIONS

	Before Test	After Test
Mass of moist soil + ring		
Mass of moist soil	$M_{To} =$	$M_{Tf} =$
Mass of dry soil + ring		
Mass of dry soil	$M_d =$	$M_d =$
Mass of moisture		
Moisture content	$w_o =$	$w_f =$
Void ratio	$e_o =$	$e_f =$
Degree of saturation	$S_o =$	$S_f =$

IV. EQUATION AND CALCULATION SPACE

$$e_o = \frac{V_o - \dfrac{M_d}{G_s \rho_w}}{\dfrac{M_d}{G_s \rho_w}}$$

ONE-DIMENSIONAL CONSOLIDATION TEST (ASTM D2435)
TIME-DEFORMATION MEASUREMENTS
LABORATORY DATA SHEET

I. GENERAL INFORMATION

Test performed by:	Date tested:
Lab partners/organization:	
Client:	Project:
Boring no.:	Recovery depth:
Recovery date:	Recovery method:
Soil description:	

II. TEST DETAILS

Load frame type/serial no.:	
Scale type/serial no./precision:	
Load no.:	Load increment, σ':
Filter paper type:	
Porous stone type and thickness:	
Machine deflection:	
Deformation indicator type and conversion factor K (if applicable):	
Notes, observations, and deviations from ASTM D2435 test standard:	

III. MEASUREMENTS AND CALCULATIONS

Date (mm/dd/yy)	Clock Time (hh:mm:ss)	Elapsed Time (hh:mm:ss)	Raw Deformation ( )	Deflection-Corrected Deformation ( )

ONE-DIMENSIONAL CONSOLIDATION TEST (ASTM D2435)
TIME-DEFORMATION PLOTTING USING THE LOG TIME METHOD

I. GENERAL INFORMATION

Data plotted by:	Date:
Lab partners/organization:	
Client:	Project:
Boring no.:	Recovery depth:
Recovery date:	Recovery method:
Soil description:	

II. TEST DETAILS

Load no.:	Load, σ':
Initial specimen height, H_o:	Deflection units:
Dial gauge conversion factor, K:	
Notes, observations, and deviations from ASTM D2435 test standard:	

III. MEASUREMENTS AND CALCULATIONS

CALCULATION SPACE:

σ':	d_{100}:
t_2:	d_2:
t_1:	d_1:
Δd:	d_o:
d_{50}:	t_{50}:
H_{D50}:	c_v:

IV. EQUATIONS

$$t_1 = t_2/4 \qquad \Delta d = d_2 - d_1 \qquad d_0 = d_1 - \Delta d \qquad d_{50} = (d_0 + d_{100})/2$$

$$H_{D50} = \frac{H_o - d_{50}(K)}{2} \text{ or } H_{D50} = \frac{H_o - d_{50}}{2} \qquad c_v = \frac{0.197(H_{D50})^2}{t_{50}}$$

EXAMPLE:

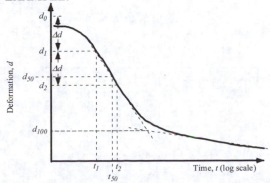

ONE-DIMENSIONAL CONSOLIDATION TEST (ASTM D2435)
TIME-DEFORMATION PLOTTING USING THE ROOT TIME METHOD

I. GENERAL INFORMATION

Data plotted by:	Date:
Lab partners/organization:	
Client:	Project:
Boring no.:	Recovery depth:
Recovery date:	Recovery method:
Soil description:	

II. TEST DETAILS

Load no.:	Load, σ':
Initial specimen height, H_o:	Deflection units:
Dial gauge conversion factor, K:	
Notes, observations, and deviations from ASTM D2435 test standard:	

III. MEASUREMENTS AND CALCULATIONS

CALCULATION SPACE:

σ':	d_0:
X:	$1.15X$:
d_{90}:	t_{90}:
d_{100}:	H_{D50}:
c_v:	

IV. EQUATIONS

$$d_{100} = d_0 + 1.11(d_{90} - d_o) \qquad c_v = \frac{0.848(H_{D50})^2}{t_{90}}$$

EXAMPLE:

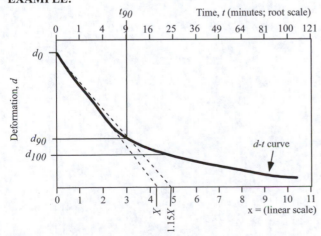

145

ONE-DIMENSIONAL CONSOLIDATION TEST (ASTM D2435)
TIME-DEFORMATION PLOTTING USING THE ROOT TIME METHOD
PLOTTING PAPER

Elapsed time t (min)

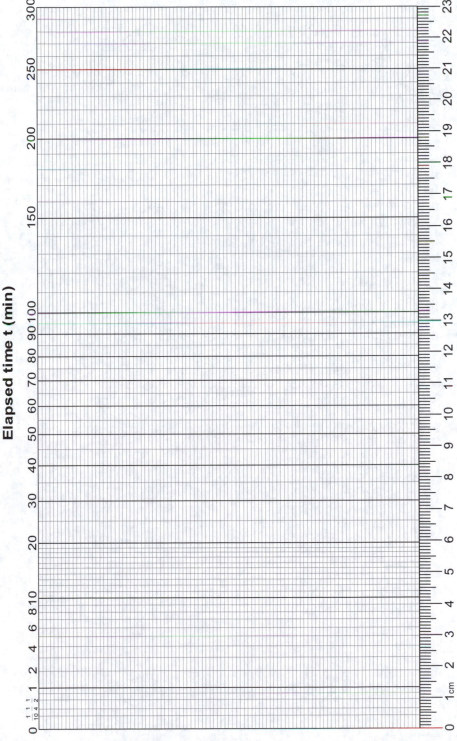

Settlement S (division)

ONE-DIMENSIONAL CONSOLIDATION TEST (ASTM D2435)
CONSTRUCTION OF $e - \log \sigma'$ CURVE

I. GENERAL INFORMATION

Plotted by:	Dates tested:
Lab partners/organization:	
Client:	Project:
Boring no.:	Recovery depth:
Soil description:	

II. TEST DETAILS

Initial specimen height, H_o:	Specimen diameter:
Initial specimen volume, V_o:	Specific gravity of soil solids, G_s:
Net dry mass of specimen, M_d:	Initial void ratio, e_o:
Deflection units:	Dial gauge conversion factor, K:
Height of solids, H_s:	
Notes, observations, and deviations from ASTM D2435 test standard:	

III. MEASUREMENTS AND CALCULATIONS

σ'	d_{100}	Δe	e

EXAMPLE:

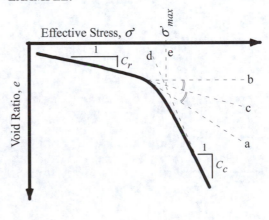

C_r:	
C_c:	
σ'_{max}:	

IV. EQUATIONS

$$e_o = \frac{V_o - \dfrac{M_d}{G_s\rho_w}}{\dfrac{M_d}{G_s\rho_w}}$$

$$H_s = \frac{H_o}{1+e_0}$$

$$\Delta e = \frac{\Delta H}{H_s} = \frac{d_{100}(K)}{H_s} \ \text{ or } \ \Delta e = \frac{\Delta H}{H_s} = \frac{d_{100}}{H_s}$$

$$e = e_0 - \Delta e$$

$$C = \frac{e_1 - e_2}{\log \sigma_2 - \log \sigma_1}$$

12. DIRECT SHEAR STRENGTH TEST OF GRANULAR SOIL

12.1. APPLICABLE ASTM STANDARDS

- ASTM D3080: Standard Test Method for Direct Shear Test of Soils Under Consolidated Drained Conditions

12.2. PURPOSE OF MEASUREMENT

Direct shear testing (ASTM D3080) provides the shear strength properties of soils under conditions of drained loading, which is required for assessing the stability of earth slopes. It is commonly used to test cohesionless soils (i.e.; sands and gravels) because of the inherent difficulties in preparing specimens of cohesionless soil for triaxial strength testing. The ASTM D3080 standard can also be applied to test cohesive soils under conditions of drained loading to derive drained shear strength parameters. However, it is seldom used for testing cohesive soils, and its use is generally limited to cohesionless soils. The drained shear strength of cohesive soils is more commonly measured using triaxial shear strength testing (ASTM D4767).

12.3. DEFINITIONS AND THEORY

With respect to shear strength, soil can be viewed as a frictional material. Consider a block of soil sheared along a failure plane as shown in Fig. 12.1. A normal force, N, is applied, and the failure plane slips when a shear force F_f is achieved. If the area of the block is A, then the normal stress and shear stress, σ and τ_f, are expressed as:

$$\sigma = \frac{N}{A} \tag{12.1}$$

and

$$\tau_f = \frac{F_f}{A}. \tag{12.2}$$

If the normal force N is increased, a higher value of F_f is required to cause the failure plane to slip. By plotting σ versus τ_f over a range of N, a Mohr-Coulomb failure envelope is defined. The intercept of the Mohr-Coulomb failure envelope is c', and the friction angle is ϕ' as shown in Fig. 12.2. Thus, τ_f can be expressed as a function of σ':

$$\tau_f = c' + \sigma' \tan \Phi. \tag{12.3}$$

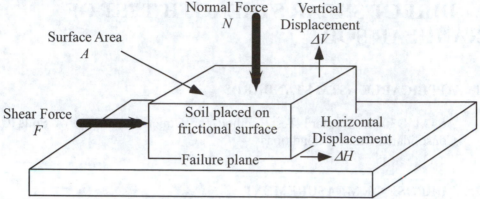

Fig. 12.1 – Schematic illustration of the direct shear test.

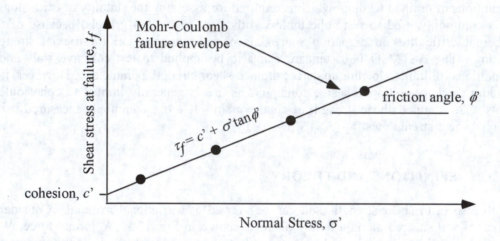

Fig. 12.2 – Mohr-Coulomb failure envelope.

The superscript (') over the c, σ, and ϕ terms in Eqn. 12.3 indicates that the strength properties measured during the test are drained properties, where pore pressure remains near zero throughout the test. This test is referred to as a direct shear test, and is often used to measure the strength of cohesionless soils where c' is approximately equal to zero. A variation of the direct shear test (ASTM D5321) is also used to estimate the frictional properties of geosynthetic materials.

Horizontal displacement *(ΔH)*, vertical displacement (*ΔV*), and F are measured during a direct shear test. Shear stress (τ) is calculated as:

$$\tau = \frac{F}{A},$$ (12.4)

and is plotted versus *ΔH* to identify τ_f. For loose soils, the $\tau - \Delta H$ curve does not exhibit a distinct peak, and τ_f is defined as τ at large strains (e.g. *ΔH* > 0.3 in.), where τ is more or less independent of *ΔH*. For dense soils, the τ - *ΔH* curve exhibits a distinct peak, and

τ_f is defined as the peak shear stress. Examples of typical $\tau - \Delta H$ curves for loose and dense soils are illustrated schematically in Fig. 12.3. If ΔV is plotted as a function of ΔH, two distinct curves result depending on whether the soil is dense or loose. For loose soils, the volume of the soil decreases during shearing. Loose soils are contractive, and the soil particles move into existing voids. As a result, the soil undergoes a net decrease in volume as ΔV decreases. For dense soils, the volume of the soil increases during shearing as the soil dilates. For dense soils, the soil particles must move up and over one another for the soil to shear, so the soil undergoes a net increase in volume as ΔV increases. This phenomenon is referred to as dilation. Examples of the contractive and dilative behavior of loose and dense soils are illustrated in Fig. 12.4.

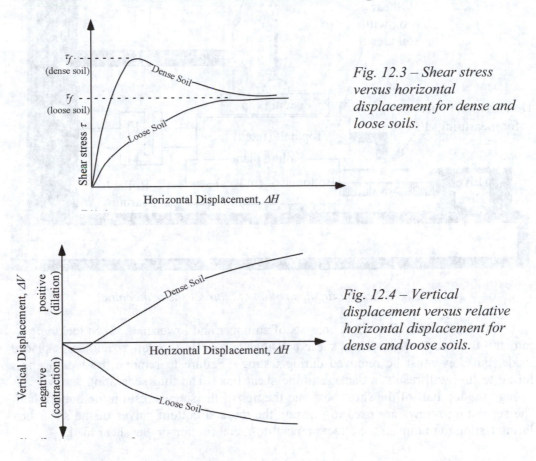

Fig. 12.3 – Shear stress versus horizontal displacement for dense and loose soils.

Fig. 12.4 – Vertical displacement versus relative horizontal displacement for dense and loose soils.

12.4.   EQUIPMENT AND MATERIALS

The following materials and equipment are required to perform direct shear testing:

- Soil;
- Shear box;
- Funnel; and
- Direct shear machine.

The direct shear machine is illustrated schematically in Fig. 12.5. Soil specimens are typically cylindrical, and are placed inside a square shear box. The minimum specimen diameter is 2.0 in., but the diameter must be at least 10 times the maximum particle size. The thickness of the specimen must be at least 0.5 in., and the minimum diameter : thickness ratio is 2:1.

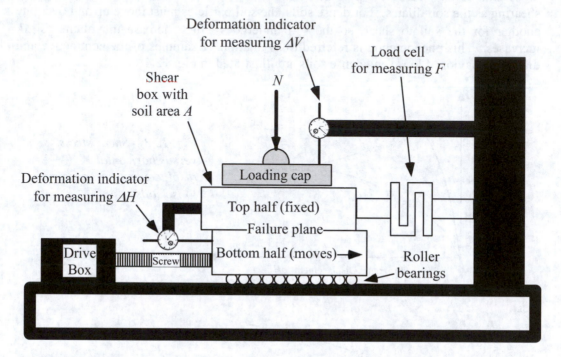

Fig. 12.5 – Schematic illustration of a direct shear machine.

The shear box (Fig. 12.6) consists of an upper and lower half. Two locking pins hold the top and bottom of the shear box together while the soil specimen is placed inside, but they must be removed during testing. Failure to remove the locking pins during testing will result in damage to the shear box. The four separating screws pass through the top half of the shear box, and the tips of the screws rest on the bottom half. The separating screws are used to separate the top and bottom halves of the shear box during testing to minimize the effect of metal-to-metal friction on the shear load.

The shear box is placed in the direct shear machine, and the test is conducted so that the plane corresponding to the boundary between the upper and lower halves of the shear box is the failure plane. The top half of the shear box is fixed against a load cell (or proving ring), and the bottom half of the box is free to move over roller bearings. Normal load, N, is usually applied to the failure plane through a loading cap on top of the specimen using dead weights. Some systems, however, use hydraulic or pneumatic pressure to apply N. Shear load, F, is applied to the failure plane by pushing against the bottom half of the box with a screw at a controlled deformation rate of between 0.0001 and 0.04 in./min. For granular soils, the strain rate can be closer to 0.04 in./min because it is not necessary to allow for excess pore pressures to dissipate. Since the top half of the box bears against the load cell, the amount of shear load carried by the failure plane and transferred from the bottom half to the top half of the box is measured directly.

Fig. 12.6 – Shear box (shown disassembled); the hole in the middle of the shear box is filled with soil during testing. Shown in photograph:

a. Loading cap;
b. Top half;
c. Bottom half;
d. Separating screws (4); and
e. Locking pins (2).

Horizontal and vertical displacement (ΔH and ΔV) are measured during the test using deformation indicators. Horizontal displacement must be measured to the nearest 0.001 in., while vertical displacement must be measured to the nearest 0.0001 in. There are several devices available for use on newer machines to measure deformation, including proximeters, LVDTs, and digital dial gauges. These devices can be calibrated to output directly in units of length, and use of these devices is relatively straightforward. However, many older machines are instrumented with analog dial gauges (Fig. 11.4). Readings are taken on an analog dial gauge in units of divisions, which are later converted to units of length using a conversion factor (e.g. 0.0001 in./division).

Most direct shear machines measure F using a load cell, but many older machines are instrumented with a proving ring. A proving ring is a large metal ring with a dial gauge positioned on the inside. As the proving ring is loaded in a radial direction, it deforms from a circular shape into an oval shape, and the amount of deformation is recorded using the dial gauge. There is a linear relationship between deformation and applied load, so the deformation observed using the dial gauge can be converted to load using a calibration constant (e.g. 30 lbs./division).

There have been many different configurations for direct shear machines manufactured and used over the years. Each design is slightly different, but all possess the same basic components shown in Fig. 12.5. A photograph of one type of direct shear machine is shown in Fig. 12.7. This particular machine is configured with a load cell for measuring F, analog dial gauges for measuring ΔH and ΔV, and a dead weight system for applying N.

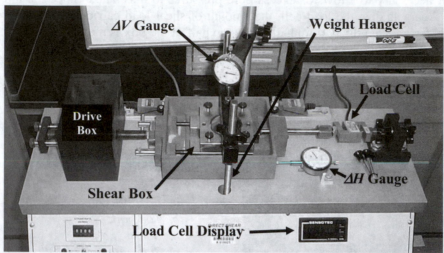

Fig. 12.6 – Photograph of a direct shear machine with analog deformation dial gauges and a load cell.

12.5. PROCEDURE[1]

The following describes the procedure for performing a direct shear test to derive a single $\sigma - \tau_f$ point for the Mohr-Coulomb failure envelope:

1) Assemble the shear box and fill it with soil provided by your laboratory instructor. Place the soil in the shear box by dropping it through the funnel from a height of approximately 1.0 in. (Fig. 12.7). Using a consistent specimen preparation method assures uniformity between specimens to obtain a linear Mohr-Coulomb failure envelope.

2) Measure the diameter of the soil specimen and calculate the area, A.

3) Place the shear box in the direct shear machine. Clamp the shear box in, and advance the screw manually so that all moving parts (screw, shear box, and load cell/proving ring) are seated snugly against one another.

4) Place the normal load, N, onto the specimen to achieve the desired level for σ (Fig. 12.8). Use Eqn. 12.1 to calculate the amount of load needed. If your machine has a dead weight hanger with a mechanical advantage, multiply the weight by the mechanical advantage to calculate the true normal force applied to the specimen. If your machine is configured with a hydraulic or pneumatic system instead of a dead weight system, ask your instructor for guidance.

---

[1] Don't forget to visit www.wiley.com/college/kalinski to view the lab demo!

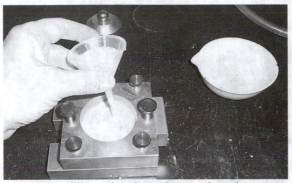

a) filling the shear box with soil   *b) filled shear box with loading cap*

Fig. 12.7 – Preparing a specimen for direct shear testing by placing soil in the shear box through a funnel.

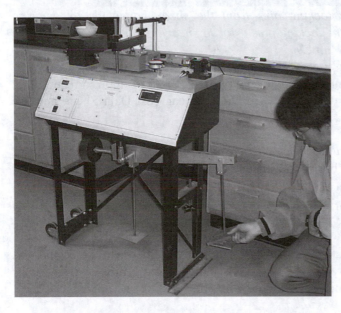

Fig. 12.8 – Applying normal load to the specimen using a dead weight system. The system shown in the photograph has a 10:1 mechanical advantage.

5) Position and zero the deformation indicators and load cell. If your machine is configured with analog dial gauges for measuring deformation, record the displacement conversion factors for the horizontal and vertical dial gauges, K_H and K_V. If your machine is configured with a proving ring instead of a load cell, record the proving ring constant, K_F.

6) REMOVE THE LOCKING PINS FROM THE SHEAR BOX (Fig. 12.9) and turn the separating screws one-quarter of a turn to separate the top and bottom halves of the shear box.

Fig. 12.9 – Removing the locking pins from the shear box prior to shearing (WARNING: Failure to remove the locking pins prior to shearing may seriously damage the shear box and direct shear machine!).

7) Begin shearing the specimen at a deformation rate $\Delta H/\Delta t$ of approximately 0.02 in/min. Record your data on the attached Direct Shear Test Data Sheet, and use additional sheets as needed.

    a. If your deformation indicators are digital dial gauges, LVDTs, or proximeters, your horizontal and vertical measurements will be ΔH and ΔV, and will be in units of length. If your machine is configured with analog dial gauges, your horizontal and vertical measurements will be G_H and G_V, and will be in units of divisions.

    b. If your force indicator is a load cell, your measurement will be F, and will be in units of force. If your force indicator is a proving ring, your measurement will be G_F, and will be in units of divisions.

    c. Record measurements frequently enough so that the peak value for F is recorded (a recording interval of $\Delta H = 0.01$ in. should be adequate). Shear the specimen until ΔH reaches 0.3 in.

    d. Vertical displacement readings will be either positive or negative, depending on whether your specimen dilates or contracts. Make sure you know the sign convention for the vertical deformation indicator so that you can accurately determine whether the specimen dilated or contracted.

8) If your deformation indicators are analog dial gauges, convert the horizontal and vertical dial gauge readings G_H and G_V to horizontal and vertical displacement, ΔH and ΔV using the following relationships:

$$\Delta H = G_H K_H \tag{12.5}$$

and

$$\Delta V = G_V K_V. \tag{12.6}$$

If your machine is configured with a proving ring, convert the proving ring reading G_F to shear force F using the following relationship:

$$F = G_F K_F. \tag{12.7}$$

9) Calculate τ for each measurement using Eqn. 12.4, and plot τ versus ΔH to identify τ_f. For dense soils, τ_f will correspond to the peak value of τ. For loose soils, τ_f will correspond to the value of τ at large strains, where τ has reached a constant value.

10) Repeat Steps 1-9 using different values for σ to derive 4 or more points for plotting of the Mohr-Coulomb failure envelope. Plot τ_f versus σ using these points to identify c' and ϕ' (Eqn. 12.3).

12.6. EXPECTED RESULTS

Typical values of c' for cohesionless soils are around zero (hence the term "cohesionless"). Typical values of ϕ' for cohesionless soil range from 30-40 degrees. Friction angle increases with increasing particle angularity, and with decreasing void ratio.

12.7. LIKELY SOURCES OF ERROR

Since ϕ' is dependent upon void ratio, the most likely source of error in the direct shear test is in specimen preparation. In the experiment described herein, each specimen is prepared in a consistent manner by dropping the soil through a funnel from a height of 1.0 in. If different preparation methods are used, however, the specimens will have different void ratios. As stated in Section 12.7, ϕ' increases with decreasing void ratio. Thus, a Mohr-Coulomb failure envelope constructed with specimens prepared using different methods will not be linear.

12.8. ADDITIONAL CONSIDERATIONS

If direct shear testing is performed as part of a Construction Quality Assurance (CQA) program for a given project to assure that design criteria are satisfied, it is important to perform the laboratory tests under conditions that are consistent with the design criteria.

Direct Shear Strength Test

For example, if design criteria state that sand used for a drainage layer is to be placed at a dry unit weight of 120 pcf, then the specimens in the laboratory should be prepared at the same dry unit weight. In this instance, extra effort would be required to measure the mass and volume of the soil in the shear box, densify the soil using either compaction or vibratory densification, and perform weight-volume calculations.

During shearing, the actual sheared area decreases as the bottom half of the shear box moves relative to the top half, which has a slight effect on τ, and σ.' To correct for this effect for a round specimen with diameter D, τ and σ would be calculated using a corrected area, A_{corr}, where:

$$A_{corr} = A \left[\frac{2}{\pi} \left(\cos^{-1} \left(\frac{\Delta H}{D} \right) - \frac{\Delta H}{D} \left(1 - \left(\frac{\Delta H}{D} \right)^2 \right) \right) \right]. \tag{12.8}$$

Application of this correction, however, will move a given τ_f-σ data point up and to the right, thus keeping it more or less on the Mohr-Coulomb failure envelope and ultimately having a negligible effect on the final calculated values for c' and ϕ.' Therefore, this correction is usually not performed, nor is it described in the ASTM D3080 test standard.

12.9. SUGGESTED EXERCISES

1) Perform 4 direct shear tests on the soil provided by your instructor using normal stresses ranging between 5 to 100 psi. Record your data on the attached Direct Shear Test Data Sheet (additional data sheets can be found on the CD-ROM that accompanies this manual).

2) Plot shear stress (τ) versus horizontal displacement (ΔH) for each test, and identify the shear stress at failure (τ_f) for each test.

3) Plot vertical displacement (ΔV) versus ΔH for each test. Did your specimens dilate or contract?

4) Plot a Mohr-Coulomb failure envelope of τ_f versus normal stress, σ for the soil. Use the Mohr-Coulomb failure envelope to calculate the cohesion (c') and friction angle (ϕ') for the soil.

DIRECT SHEAR TEST (ASTM D3080)
LABORATORY DATA SHEET

I. GENERAL INFORMATION

Tested by:	Date tested:
Lab partners/organization:	
Client:	Project:
Boring no.:	Recovery depth:
Recovery date:	Recovery method:
Soil description:	

II. TEST DETAILS

Sample diameter:	Sample area, A:
Normal force, N:	Normal stress, σ:
Deformation rate:	Deformation indicator type:
Shear force measurement instrument type:	
Horizontal dial gauge conversion factor, K_H:	
Vertical dial gauge conversion factor, K_V:	
Proving ring dial gauge conversion factor, K_F:	
Notes, observations, and deviations from ASTM D3080 test standard:	

III. MEASUREMENTS AND CALCULATIONS

Horizontal Deformation Reading (G_V)	Vertical Deformation Reading (G_H)	Force Reading (G_F)	Horizontal Displacement (ΔH)	Vertical Displacement (ΔV)	Shear Force (F)	Shear Stress (τ)

Shear strength (τ_f):	

DIRECT SHEAR TEST (ASTM D3080)
LABORATORY DATA SHEET

I. GENERAL INFORMATION

Tested by:	Date tested:
Lab partners/organization:	
Client:	Project:
Boring no.:	Recovery depth:
Recovery date:	Recovery method:
Soil description:	

II. TEST DETAILS

Sample diameter:	Sample area, A:
Normal force, N:	Normal stress, σ:
Deformation rate:	Deformation indicator type:
Shear force measurement instrument type:	
Horizontal dial gauge conversion factor, K_H:	
Vertical dial gauge conversion factor, K_V:	
Proving ring dial gauge conversion factor, K_F:	
Notes, observations, and deviations from ASTM D3080 test standard:	

III. MEASUREMENTS AND CALCULATIONS

Horizontal Deformation Reading (G_V)	Vertical Deformation Reading (G_H)	Force Reading (G_F)	Horizontal Displacement (ΔH)	Vertical Displacement (ΔV)	Shear Force (F)	Shear Stress (τ)

Shear strength (τ_f):

DIRECT SHEAR TEST (ASTM D3080)
LABORATORY DATA SHEET

I. GENERAL INFORMATION

Tested by:	Date tested:
Lab partners/organization:	
Client:	Project:
Boring no.:	Recovery depth:
Recovery date:	Recovery method:
Soil description:	

II. TEST DETAILS

Sample diameter:	Sample area, A:
Normal force, N:	Normal stress, σ.
Deformation rate:	Deformation indicator type:
Shear force measurement instrument type:	
Horizontal dial gauge conversion factor, K_H:	
Vertical dial gauge conversion factor, K_V:	
Proving ring dial gauge conversion factor, K_F:	
Notes, observations, and deviations from ASTM D3080 test standard:	

III. MEASUREMENTS AND CALCULATIONS

Horizontal Deformation Reading (G_V)	Vertical Deformation Reading (G_H)	Force Reading (G_F)	Horizontal Displacement (ΔH)	Vertical Displacement (ΔV)	Shear Force (F)	Shear Stress (τ)

Shear strength (τ_f):	

DIRECT SHEAR TEST (ASTM D3080)
LABORATORY DATA SHEET

I. GENERAL INFORMATION

Tested by:	Date tested:
Lab partners/organization:	
Client:	Project:
Boring no.:	Recovery depth:
Recovery date:	Recovery method:
Soil description:	

II. TEST DETAILS

Sample diameter:	Sample area, A:
Normal force, N:	Normal stress, σ:
Deformation rate:	Deformation indicator type:
Shear force measurement instrument type:	
Horizontal dial gauge conversion factor, K_H:	
Vertical dial gauge conversion factor, K_V:	
Proving ring dial gauge conversion factor, K_F:	
Notes, observations, and deviations from ASTM D3080 test standard:	

III. MEASUREMENTS AND CALCULATIONS

Horizontal Deformation Reading (G_V)	Vertical Deformation Reading (G_H)	Force Reading (G_F)	Horizontal Displacement (ΔH)	Vertical Displacement (ΔV)	Shear Force (F)	Shear Stress (τ)

Shear strength (τ_f):

13. UNCONFINED COMPRESSIVE STRENGTH TEST

13.1. APPLICABLE ASTM STANDARDS

- ASTM D2166: Standard Test Method for Unconfined Compressive Strength of Cohesive Soil

13.2. PURPOSE OF MEASUREMENT

Unconfined compressive strength testing provides a quick and simple means to measure the unconfined compressive strength (q_u) and undrained shear strength (s_u) of normally consolidated and slightly overconsolidated cylindrical specimens of cohesive soil. This information is used to estimate the bearing capacity of spread footings and other structures when placed on deposits of cohesive soil.

13.3. DEFINITIONS AND THEORY

With respect to shear strength, cohesive soil can fail under conditions of rapid loading where excess pore pressures do not have time to dissipate. Under these conditions, the state of stress in an element of soil can be illustrated in terms of a Mohr circle, with minor and major total principal stress σ_3 and σ_{1f}, respectively. If identical specimens of cohesive soil are subjected to different states of stress and rapidly loaded to failure without excess pore pressure dissipation, the Mohr circles of each specimen possess the same diameter, thus producing a "total stress envelope" with a friction angle of zero, and cohesion equal to the undrained shear strength, s_u (Fig. 13.1). It is important to note, however, that if pore pressure is measured within each specimen during shearing and total stresses are converted to effective stresses, each Mohr circle overlaps one another and is tangent to the effective stress envelope with an effective cohesion c' and effective friction angle ϕ'. This illustrates an important point regarding the strength of soil: even under rapid undrained loading, the strength of soil is still controlled by effective stress!

To obtain information for defining a total stress envelope, undisturbed specimens are often strength tested using the unconsolidated-undrained (UU) triaxial test (a.k.a. Q-test, ASTM D2850), where the specimen is placed in a pressurized triaxial cell with σ_3 equal to the cell pressure, and σ_1 equal to the cell pressure plus a deviator stress applied to the top of the specimen with a piston. The UU triaxial test requires the use of a triaxial cell, where the soil specimen is sealed in a latex membrane, placed in a pressurized, water-filled triaxial cell, and tested. For overconsolidated soil specimens with fissures that can act as preferential planes of weakness, ASTM D2850 is a preferable method that will prevent the specimen from failing along these preexisting planes to provide an accurate, representative measure of the in situ strength of the specimen. However, for normally consolidated or slightly overconsolidated specimens, the specimen will not suffer from the effects of fissuring, so use of a triaxial cell to achieve a nonzero σ_3 is not necessary. Under these conditions, σ_1 can be applied to an undisturbed specimen by loading it in a load frame under a constant strain rate, and a single Mohr circle with $\sigma_3 = 0$ can be plotted to estimate s_u. This test is referred to as the unconfined compressive strength test (ASTM D2166). It is a faster, simpler alternative to the UU triaxial test that does not require the use of a triaxial cell, latex membrane, or pressure source. As shown in Fig. 13.2, s_u is defined as the intercept of the total stress failure envelope, and is half of the diameter of the Mohr circle. The unconfined compressive strength, q_u, is defined as σ_1 at failure. By inspection, s_u is equal to one half of q_u.

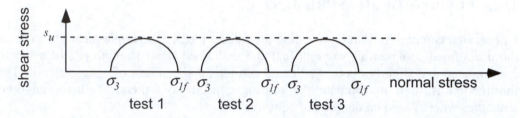

Fig. 13.1 – Total stress Mohr-Coulomb failure envelope.

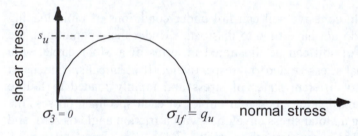

Fig. 13.2 – Mohr circle from an unconfined compressive strength test.

A typical configuration for the unconfined compressive strength testing is shown in Fig. 13.3. Axial deformation, ΔL, is measured using a deformation indicator, and applied load, P, is measured using a load cell. Axial strain, ε_1, is expressed as:

$$\varepsilon_1 = \Delta L/L_o, \tag{13.1}$$

where is L_o is the initial length of the specimen. As the specimen deforms, its cross-sectional area increases, and the corrected area, A, is expressed as:

$$A = A_o/(1-\varepsilon_1), \tag{13.2}$$

where A_o is the initial cross-sectional area of the specimen. The major principal stress, σ_1, is expressed as:

$$\sigma_1 = P/A. \tag{13.3}$$

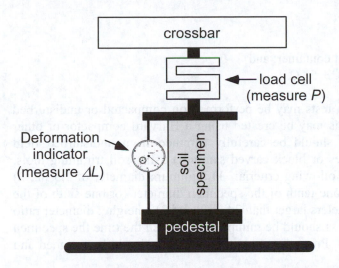

Fig. 13.3 – *Typical configuration for an unconfined compressive strength test.*

Unconfined compressive strength testing is performed by straining the specimen at a constant axial strain rate of between 0.5-2.0%/min. Some systems apply load using a moving crossbar and fixed pedestal, while others use a fixed crossbar and a moving pedestal. During the test, σ_1 is plotted versus ε_1 to identify q_u (Fig. 13.4). For stiff clays, q_u is defined as the peak of the $\sigma_1 - \varepsilon_1$ curve. For soft clays, q_u is defined as σ_1 at a strain level of 15%.

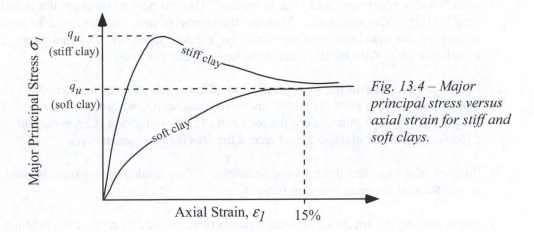

Fig. 13.4 – *Major principal stress versus axial strain for stiff and soft clays.*

13.4. EQUIPMENT AND MATERIALS

The following equipment and materials are required to perform unconfined compressive strength testing:

- Right-circular cylindrical specimen of cohesive soil;
- load frame;
- deformation indicator graduated to 0.001 in.;
- load cell or proving ring;

- scale with precision of 0.01 g;
- calipers;
- oven-safe moisture content container; and
- soil drying oven set at $110^\circ \pm 5\,^\circ\mathrm{C}$.

Unconfined compressive strength tests may be performed on compacted or undisturbed specimens. Compacted specimens may be created using a Harvard compactor or other device. Undisturbed specimens should be carefully trimmed from undisturbed field samples (e.g. Shelby tube samples or block carved samples) using soil trimming tools. Test specimens must satisfy the following criteria: 1) minimum diameter of 1.3 in., 2) maximum particle size less than one-tenth of the specimen diameter (or one-sixth of the diameter for specimens with diameters larger than 2.8 in.), and 3) a height : diameter ratio between 2.0 and 2.5. Moisture loss should be minimized between the time the specimen is prepared and when it is tested. Prior to testing, the specimen should be weighed and measured.

13.5.   PROCEDURE[1]

The procedure for performing an unconfined compressive strength test on a cylindrical specimen of cohesive soil is as follows:

1) Obtain a test specimen from your instructor. Use calipers to measure the initial length (L_o) of the specimen. Measure the diameter near the top, middle, and bottom of the specimen, and calculate the average diameter (D_o) and average initial area (A_o). Also measure the moist mass of the specimen (M).

2) Place the specimen in the load frame, and advance the pedestal (or crossbar) so that all the moving parts (pedestal, specimen, load cell, and crossbar) are seated snugly against each other. Zero the load cell. If a proving ring is used instead of a load cell, zero the dial gauge and record the proving ring constant K_P.

3) Position and zero the deformation indicator. If an analog dial gauge is used, record the dial gauge conversion factor K_L.

4) Begin loading the specimen at a strain rate between 0.5-2.0%/min. Take readings frequently enough to fully define the peak of the curve during the test. Record your data on the Unconfined Compressive Strength Test Data Sheet, and use additional sheets as needed. Load the specimen until $\varepsilon_l = 15\%$.

5) If your deformation indicator is a digital dial gauge, proximeter, or LVDT, your reading will be ΔL, and will be in units of length. If your deformation indicator is an analog dial gauge, your reading will be G_L, and will be in units of divisions. For analog dial gauges, ΔL is calculated as:

---

[1] Don't forget to visit www.wiley.com/college/kalinski to view the lab demo!

$$\Delta L = G_L K_L, \tag{13.4}$$

6) If your load frame is configured with a load cell, your reading will be P, and will be in units of force. If your load frame is configured with a proving ring instead of a load cell, your reading will be G_P, and will be in units of divisions. For proving rings, P is calculated as:

$$P = G_P K_P. \tag{13.5}$$

7) Plot σ_l versus ε_l and identify q_u as either 1) the peak value of σ_l or 2) σ_l at $\varepsilon_l = 15\%$.

8) Place the specimen in a soil drying oven overnight and obtain the dry weight of the specimen, M_s, for weight-volume calculations.

13.6. EXPECTED RESULTS

Unconfined compressive strength of fine-grained soils may range from a few psi for soft, normally consolidated clays, to over 50 psi for dry compacted specimens. For stiffer specimens, a failure plane may be apparent within the specimen, oriented at an angle of approximately 45 degrees (Fig. 13.5). Softer specimens are less likely to exhibit a distinct failure plane, and are more likely to demonstrate "barreling" behavior.

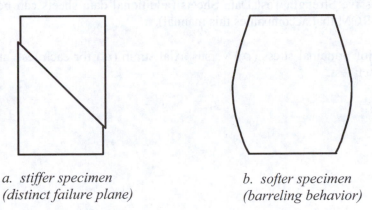

a. stiffer specimen
(distinct failure plane)

b. softer specimen
(barreling behavior)

Fig. 13.5 – Typical appearance of failed specimens after unconfined compressive strength testing.

13.7. LIKELY SOURCES OF ERROR

The unconfined compressive strength test is appropriate for normally consolidated or slightly overconsolidated undisturbed specimens, or for compacted specimens of fine-grained soil. When these specimens are tested without confinement, the failure plane will develop within the specimen. However, highly overconsolidated specimens of undisturbed soil may possess cracks and fissures. If these specimens are tested without confinement, failure may occur along one of these preexisting surfaces. In this case, the strength of the soil will be underestimated. For highly overconsolidated soils, undrained shear strength should be measured using the unconsolidated-undrained (UU) type triaxial test (ASTM D2850), which is described in Chapter 14.

13.8.   ADDITIONAL CONSIDERATIONS

Unconfined compressive strength testing provides an estimate for the undrained shear strength of fine-grained soil, which describes how soil will behave under short-term conditions of rapid loading when excess pore pressures are not allowed to dissipate. This is most commonly used to assess the load bearing capacity of soil. However, it is often necessary to estimate the shear strength under conditions of long-term loading when excess pore pressures do not develop. One common example is in the assessment of earth slope stability. Under these conditions, it is necessary to estimate the drained strength parameters using triaxial strength testing (ASTM D4767).

13.9.   SUGGESTED EXERCISES

1) Perform unconfined compressive strength testing on two specimens of fine-grained soil provided by your instructor. Use the attached Unconfined Compressive Strength Test Data Sheets (additional data sheets can be found on the CD-ROM that accompanies this manual).

2) Plot major principal stress (σ_l) versus axial strain (ε_l) for each test, and identify q_u for each test.

UNCONFINED COMPRESSIVE STRENGTH TEST (ASTM D2166)
LABORATORY DATA SHEET

I. GENERAL INFORMATION

Tested by:	Date tested:
Lab partners/organization:	
Client:	Project:
Boring no.:	Recovery depth:
Recovery date:	Recovery method:
Soil description:	

II. TEST DETAILS

Initial specimen diameter, D_o:	Initial specimen area, A_o:	
Initial specimen length, L_o:	Initial specimen volume, V_o:	
Moist mass of specimen, M:	Dry mass of specimen, M_s:	
Moisture content, w:	Total unit weight, γ:	Dry unit weight, γ_d:
Specimen preparation method:		
Deformation indicator type:	Axial strain rate, $\Delta\varepsilon_l/\Delta t$:	
Deformation dial gauge conversion factor, K_L:		
Force measurement instrument type:		
Proving ring dial gauge conversion factor, K_P:		
Notes, observations, and deviations from ASTM D2166 test standard:		

III. MEASUREMENTS AND CALCULATIONS

Deformation Reading (G_L)	Axial Deformation (ΔL)	Load Reading (G_P)	Axial Load (P)	Axial Strain (ε_l)	Corrected Area (A)	Axial Stress (σ_l)

EQUATIONS:

$$\varepsilon_l = \Delta L/L_o$$

$$A = A_o/(1-\varepsilon_l)$$

$$\sigma_l = P/A$$

$$\Delta L = G_L K_L$$

$$P = G_P K_P$$

$$s_u = q_u/2$$

Unconfined compressive strength, q_u:	
Undrained shear strength, s_u:	

UNCONFINED COMPRESSIVE STRENGTH TEST (ASTM D2166)
LABORATORY DATA SHEET

I. GENERAL INFORMATION

Tested by:	Date tested:
Lab partners/organization:	
Client:	Project:
Boring no.:	Recovery depth:
Recovery date:	Recovery method:
Soil description:	

II. TEST DETAILS

Initial specimen diameter, D_o:	Initial specimen area, A_o:
Initial specimen length, L_o:	Initial specimen volume, V_o:
Moist mass of specimen, M:	Dry mass of specimen, M_s:
Moisture content, w:	Total unit weight, γ:	Dry unit weight, γ_d:
Specimen preparation method:		
Deformation indicator type:	Axial strain rate, $\Delta\varepsilon_l/\Delta t$:
Deformation dial gauge conversion factor, K_L:	
Force measurement instrument type:	
Proving ring dial gauge conversion factor, K_P:	
Notes, observations, and deviations from ASTM D2166 test standard:	

III. MEASUREMENTS AND CALCULATIONS

EQUATIONS:

Deformation Reading (G_L)	Axial Deformation (ΔL)	Load Reading (G_P)	Axial Load (P)	Axial Strain (ε_l)	Corrected Area (A)	Axial Stress (σ_l)

$\varepsilon_l = \Delta L/L_o$

$A = A_o/(1-\varepsilon_l)$

$\sigma_l = P/A$

$\Delta L = G_L K_L$

$P = G_P K_P$

$s_u = q_u/2$

Unconfined compressive strength, q_u:	
Undrained shear strength, s_u:	

14. UNCONSOLIDATED UNDRAINED TRIAXIAL STRENGTH TESTING

14.1. APPLICABLE ASTM STANDARDS

- ASTM D2850: Standard Test Method for Unconsolidated Undrained Triaxial Compression Test on Cohesive Soils

14.2. PURPOSE OF MEASUREMENT

The Unconconsolidated Undrained (UU) triaxial strength test provides a means to measure the undrained shear strength (s_u) of overconsolidated cylindrical specimens of cohesive soil. This information is used to estimate the bearing capacity of spread footings and other structures when placed on deposits of cohesive soil. The UU test is also referred to as the Q test because it is a relatively fast (quick) test.

14.3. DEFINITIONS AND THEORY

With respect to shear strength, cohesive soil can fail under conditions of rapid loading, where excess pore pressures do not have time to dissipate. Under these conditions, the state of stress in an element of soil can be represented by a Mohr circle. At failure, the minor and major total principal stresses are σ_3 and σ_{lf}, respectively. If identical specimens of cohesive soil are subjected to different states of stress and rapidly loaded to failure without allowing excess pore pressure dissipation, the Mohr circles of each specimen possess the same diameter, thus producing a "total stress envelope" with a friction angle of zero, and cohesion equal to the undrained shear strength, s_u (Fig. 14.1). It is important to note, however, that if pore pressure is measured within each specimen during shearing and total stresses are converted to effective stresses, each Mohr circle overlaps one another and is tangent to the effective stress envelope with an effective cohesion c' and effective friction angle ϕ'. This illustrates an important point regarding the strength of soil: even under rapid undrained loading, the strength of soil is still controlled by effective stress!

To obtain information for defining a total stress envelope, undisturbed specimens are often strength tested using the unconsolidated-undrained (UU) triaxial test (a.k.a. Q test, ASTM D2850), where the specimen is placed in a pressurized triaxial cell with σ_3 equal to the cell pressure, and σ_l equal to the cell pressure plus a deviator stress applied to the top of the specimen with a piston. The UU triaxial test requires the use of a triaxial cell, where the soil specimen is sealed in a latex membrane, placed in a pressurized, water-filled triaxial cell, and tested. For normally consolidated or slightly overconsolidated soils, the unconfined compression test (Chapter 13) is a simpler alternative to the UU triaxial test. However, for overconsolidated soil specimens with fissures that can act as preferential planes of weakness, ASTM D2850 is a preferable method that will prevent the specimen from failing along these preexisting planes to provide an accurate, representative measure of the in situ strength of the specimen.

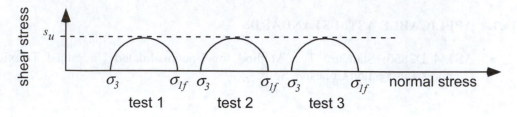

Fig. 14.1 – Total stress Mohr-Coulomb failure envelope.

A typical configuration for the UU triaxial test is shown in Fig. 14.2. The soil specimen is placed between the base and cap, and sealed using a latex membrane and O-rings. The sealed specimen is placed in a water-filled triaxial pressure cell. The cell wall is typically constructed of clear acrylic plastic, while the pedestal and top are typically metal. Cell walls are often configured with metal belts to provide extra resistance against rupturing under pressure. The base is fixed to the pedestal. A piston passes through the top of the cell, which transfers load through the cap to the specimen. Testing is performed by applying load to the piston at a constant strain rate.

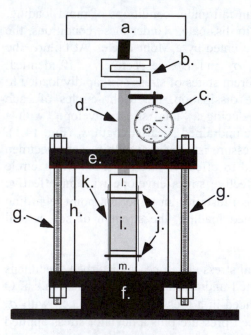

Fig. 14.2 – Experimental configuration for UU triaxial testing. Parts include:

a. crossbar;
b. load cell;
c. deformation indicator;
d. piston;
e. top;
f. pedestal;
g. cell bars;
h. cell wall;
i. soil specimen;
j. O-rings;
k. latex membrane;
l. cap; and
m. base.

Axial deformation, ΔL, is measured using the deformation indicator, and deviator load, P, is measured using the load cell. Axial strain, ε_l, is expressed as:

$$\varepsilon_l = \Delta L/L_o, \qquad (14.1)$$

where L_o is the initial length of the specimen. As the specimen deforms, its cross-sectional area increases, and the corrected area, A, is expressed as:

$$A = A_o/(1-\varepsilon_1), \tag{14.2}$$

where A_o is the initial cross-sectional area of the specimen. The deviator stress, $\Delta\sigma$, is expressed as:

$$\Delta\sigma = P/A. \tag{14.3}$$

The minor principal stress, σ_3, is equal to the cell pressure, and the major principal stress, σ_1, is equal to:

$$\sigma_1 = \sigma_3 + \Delta\sigma. \tag{14.4}$$

UU triaxial testing is performed by straining the specimen at a constant axial strain rate of between 0.3 and 1.0%/min. Plastic materials should be strained at a rate closer to 1.0%/min., while brittle materials should be strained at a rate closer to 0.3%/min. Some systems apply load using a moving crossbar and fixed pedestal, while others use a fixed crossbar and a moving pedestal. Major principal stress is plotted versus ε_1 to identify the major principal stress at failure, σ_{1f} (Fig. 14.3). For stiffer clays, σ_{1f} is defined as the peak of the $\sigma_1 - \varepsilon_1$ curve. For softer clays, σ_{1f} is defined as σ_1 at an axial strain of 15%.

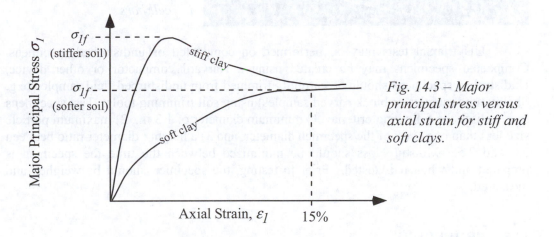

Fig. 14.3 – Major principal stress versus axial strain for stiff and soft clays.

14.4. EQUIPMENT AND MATERIALS

The following equipment and materials are required to perform UU triaxial testing:

- Right-circular cylindrical specimen of cohesive soil;
- load frame;
- pressure system and water source;

- triaxial cell;
- 2 O-rings;
- latex membrane;
- membrane stretcher;
- vacuum grease;
- deformation indicator graduated to 0.001 in.;
- load cell or proving ring;
- scale with precision of 0.01 g;
- calipers;
- oven-safe moisture content container; and
- soil drying oven set at $110^\circ \pm 5^\circ$ C.

Figure 14.4 is a disassembled triaxial cell illustrating each of the individual components.

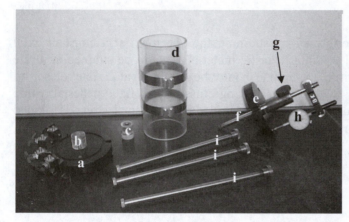

Fig. 14.4 – Disassembled triaxial cell. Parts include:
a. pedestal;
b. base;
c. cap;
d. cell wall;
e. top;
f. piston;
g. locking screw;
h. deformation indicator; and
i. cell bars

UU triaxial tests may be performed on compacted or undisturbed specimens. Compacted specimens may be created using a Harvard compactor or other device. Undisturbed specimens should be carefully trimmed from undisturbed field samples (e.g. Shelby tube samples or block carved samples) using soil trimming tools. Test specimens must satisfy the following criteria: 1) minimum diameter of 1.3 in., 2) maximum particle size less than one-sixth of the specimen diameter, and 3) a height : diameter ratio between 2.0 and 2.5. Moisture loss should be minimized between the time the specimen is prepared and when it is tested. Prior to testing, the specimen should be weighed and measured.

14.5. PROCEDURE[1]

The procedure for performing a UU triaxial test on a cylindrical specimen of cohesive soil is as follows:

1) Obtain a soil specimen from your instructor. Use calipers to measure the initial length (L_o) of the specimen. Measure the diameter near the top, middle, and

---

[1] Don't forget to visit www.wiley.com/college/kalinski to view the lab demo!

bottom of the specimen, and calculate the average diameter (D_o) and average initial area (A_o). Also measure the moist mass of the specimen (M).

2) Apply a light coating of vacuum grease to the perimeter of the base and cap to help create a waterproof seal (Fig. 14.5).

Fig. 14.5 – Applying vacuum grease to the base.

3) Place the soil specimen on the base, and place the cap on top of the specimen (Fig. 14.6). Make sure that the piston hole in the cap faces up.

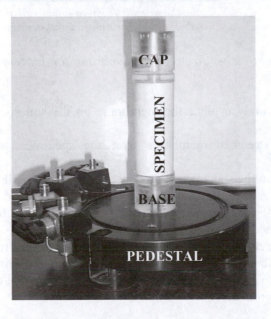

Fig. 14.6 – Specimen in place with base and cap (NOTE: a section of white PVC pipe is used in the photographs for demonstration purposes).

4) Place the membrane and two O-rings on the membrane stretcher, and apply light vacuum to the membrane stretcher tube to pull the membrane towards the inside wall of the membrane stretcher (Fig. 14.7).

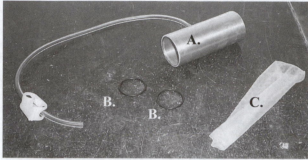

A. membrane stretcher
B. O-rings
C. latex membrane

a. membrane stretcher, O-rings, and membrane

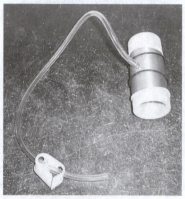

b. membrane and O-rings on membrane stretcher

c. membrane pulled against inside wall of membrane stretcher using a light vacuum

Fig. 14.7 – Using membrane stretcher to prepare membrane for placement on specimen.

5) The following steps describe how to place the membrane on the soil specimen (Fig. 14.8):

   a. Carefully lower the stretched membrane over the specimen without touching the specimen.
   b. Center the membrane on the specimen and release the vacuum to allow the membrane to constrict around the specimen.
   c. Gently pull the ends of the membrane over the base and cap so that the membrane surrounds the base, specimen, and cap without wrinkles.
   d. With the membrane stretcher still around the specimen, carefully roll the O-rings onto the membrane where the membrane contacts the base and cap. If the base and cap are machined with grooves, make sure that the O-rings are seated in the grooves.

Fig. 14.8 – Placing the membrane on the specimen.

6) The following steps describe how to assemble the triaxial cell (Fig. 14.9):

   a. Place a light coating of vacuum grease on the O-rings in the pedestal and top.

   b. Place the cell wall on the pedestal, and make sure the pedestal and cell wall are properly seated against one another.

   c. Place the top on the cell wall, and make sure the cell wall and top are properly seated against one another.

   d. Slide the piston down into the hole in the cap. The tip of the piston should be far enough into the hole to prevent the specimen from tipping when the triaxial cell is moved, but should not be applying any load to the cap. Once in position, lock the piston in place by turning the locking screw in the top.

   e. Tighten each of the three cell bars a little bit at a time, alternating between bars to assure an intimate seal between the pedestal, cell wall, and top.

c. *Placing vacuum grease on the O-ring in the pedestal.*

a. *Cell wall and top in position, with piston seated in the cap.*

b. *Top of triaxial cell with cell bars in place. Vent valve is also shown.*

Fig. 14.9 – Assembling the triaxial cell.

7) Open the vent valve in the top of the triaxial cell, and begin filling the triaxial cell with water from the pedestal valve. Shut off all valves to the triaxial cell when water emerges from the vent valve.

8) Position the triaxial cell in the load frame with the deformation indicator and load cell (Fig. 14.10).

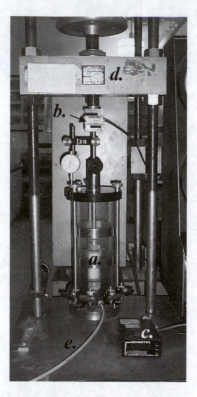

Fig. 14.10 – Positioning the triaxial cell in the load frame. Components shown include:
  a. *triaxial cell;*
  b. *load cell;*
  c. *load cell display;*
  d. *crossbar; and*
  e. *water line to controlled pressure source.*

9) Apply the desired cell pressure σ_3 to the cell through the bottom valve. You will know the specimen is under pressure when the membrane appears to be in intimate contact with the specimen.

10) Release the piston by loosening the locking screw in the top of the triaxial cell, and zero the load cell. If a proving ring is used instead of a load cell, zero the dial gauge and record the proving ring constant K_P.

11) Zero the deformation indicator. If an analog dial gauge is used, record the dial gauge conversion factor K_L.

12) Manually advance the piston until the tip of the piston is seated against the cap. You will know it is seated when the load cell begins to indicate a slight load. Once the load cell indicates a slight load, stop advancing the piston.

13) Begin loading the specimen at a strain rate between 0.3-1.0%/min. ASTM D2850 suggests that initial readings be taken at 0.1%, 0.2%, 0.3%, 0.4%, and 0.5%, 1.0%, 1.5%, 2.0%, 2.5%, and 3.0%. After that, readings should be taken at a strain interval of 1.0%. However, it may be necessary to take readings more frequently to accurately identify the peak applied load. Record your data on the Unconsolidated Undrained Triaxial Test Data Sheet, using additional sheets as needed. Load the specimen until $\varepsilon_l = 15\%$.

14) If your deformation indicator is a digital dial gauge, proximeter, or LVDT, your reading will be ΔL, and will be in units of length. If your deformation indicator is an analog dial gauge, your reading will be G_L, and will be in units of divisions. For analog dial gauges, ΔL is calculated as:

$$\Delta L = G_L K_L, \qquad (14.5)$$

15) If your load frame is configured with a load cell, your reading will be P, and will be in units of force. If your load frame is configured with a proving ring instead of a load cell, your reading will be G_P, and will be in units of divisions. For proving rings, P is calculated as:

$$P = G_P K_P. \qquad (14.6)$$

16) Plot $\Delta\sigma$ versus ε_1. Identify the deviator stress at failure, $\Delta\sigma_f$, as either 1) the peak value of $\Delta\sigma$ or 2) $\Delta\sigma$ at $\varepsilon_l = 15\%$. Calculate σ_{1f} as follows:

$$\sigma_{1f} = \sigma_3 + \Delta\sigma_f. \qquad (14.7)$$

17) Place the specimen in a soil drying oven overnight and obtain the dry weight of the specimen, M_s, for weight-volume calculations.

18) Repeat Steps 1-17 for 3 or more additional specimens tested over a range of σ_3. Plot the Mohr circles for each specimen to define the Mohr-Coulomb failure envelope and s_u.

14.6.   EXPECTED RESULTS

Undrained shear strength of fine-grained soils may range from a few psi for soft, normally consolidated clays, to over 50 psi for dry compacted specimens. For stiffer specimens, a failure plane may be apparent within the specimen, oriented at an angle of approximately 45 degrees (Fig. 14.11). Softer specimens are more likely to demonstrate "barreling" behavior.

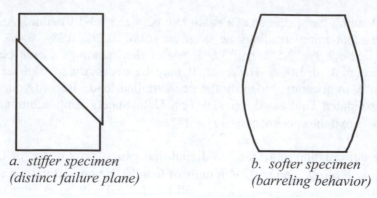

a. *stiffer specimen*
(distinct failure plane)

b. *softer specimen*
(barreling behavior)

Fig. 14.11 – Typical appearance of failed specimens after unconsolidated undrained triaxial testing.

14.7. LIKELY SOURCES OF ERROR

During UU triaxial testing, a specimen is placed under a confining stress of σ_3 without drainage. For fully saturated specimens ($S = 100\%$) loaded without drainage, all of the load is carried by the pore water, and the effective stress within the specimen remains the same regardless of σ_3. As a result, all of specimens possess the same strength, and the Mohr-Coulomb failure envelope is horizontal. However, partially saturated specimens can consolidate without drainage, so strength increases with increasing σ_3. In this case, the Mohr-Coulomb failure envelope is slightly curved (Fig. 14.12).

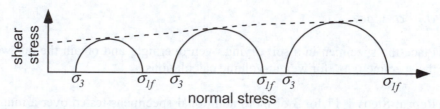

Fig. 14.12 – Mohr-Coulomb failure envelope for partially saturated specimens.

Regardless of whether specimens are partially saturated or fully saturated, the most likely source of error in the UU triaxial test stems from the fact that it is difficult to achieve perfect uniformity between test specimens. Different specimens will possess different strengths, no matter how similar they may be in their origin or preparation methods. As a result, it may be difficult to obtain a perfectly horizontal Mohr-Coulomb failure envelope.

14.8. ADDITIONAL CONSIDERATIONS

UU triaxial testing provides an estimate for the undrained shear strength of fine-grained soil, which describes how soil will behave under short-term conditions of rapid loading when excess pore pressures are not allowed to dissipate. This is most commonly used to assess the load bearing capacity of soil. However, it is often necessary to estimate the shear strength under conditions of long-term loading when excess pore pressures do not develop. One common example is in the assessment of earth slope stability. Under these conditions, it is necessary to estimate the drained strength parameters using triaxial strength testing (ASTM D4767).

Triaxial strength tests performed using ASTM D4767 are conducted in two stages. During the first stage, a saturated specimen is placed under a cell pressure of σ_3. Drainage ports in the base and cap allow excess pore pressure to dissipate as the specimen consolidates. The first stage is referred to as the consolidation stage, and is denoted by the letter C. During the second stage, the specimen is loaded to failure. The second stage is referred to as the shearing stage. The shearing stage may be performed under undrained conditions by closing the drainage ports and measuring pore pressure within the specimen (the CŪ test), or under drained conditions by opening the drainage ports (the CD test). Both of these tests provide information regarding the effective stress failure envelope (c' and ϕ') for the soil. The CŪ test can be performed rapidly, while the CD test must be performed slowly to prevent the development of excess pore pressures. In practice, the CŪ test is more common than the CD test.

Table 14.1 summarizes each of the three triaxial tests. The UU test is also called the Q test because it is quick. The CD test is also called the S test because it is slow. The CŪ test is also called the $\overline{R}$ test because R is between Q and S in the alphabet. The bar over the U and R in the CŪ ($\overline{R}$) test denotes that pore pressure is measured during the test.

Table 14.1 – Summary of triaxial strength tests for soil.

Test	ASTM Standard	Consolidation Stage	Shearing Stage	Strain Rate (%/min.)	Strength Parameters Derived
UU (Q)	D2850	unconsolidated	undrained	0.3-1.0	s_u
CŪ ($\overline{R}$)	D4767	consolidated	undrained with pore pressure measurement	0.1-0.5	c' and ϕ'
CD (S)	D4767	consolidated	drained	0.01-0.05	c' and ϕ'

In most cases, the effect of membrane stiffness on $\Delta\sigma$ is assumed to be negligible. However, ASTM D2850 describes a procedure for accounting for membrane stiffness based on some simplifying assumptions.

14.9. SUGGESTED EXERCISES

1) Perform two UU triaxial tests on specimens of fine-grained soil provided by your instructor. Use the attached Unconsolidated Undrained Triaxial Test Data Sheets (additional data sheets can be found on the CD-ROM that accompanies this manual).

2) Plot deviator stress ($\Delta\sigma$) versus axial strain (ε_l) for your tests, identify the deviator stresses at failure ($\Delta\sigma_f$), and calculate the major principal stresses at failure (σ_{1f}).

3) Use your data from the two tests to plot a Mohr-Coulomb failure envelope, and calculate the undrained shear strength (s_u) of the soil.

UNCONSOLIDATED-UNDRAINED TRIAXIAL TEST (ASTM D2850)
LABORATORY DATA SHEET

I. GENERAL INFORMATION

Tested by:	Date tested:
Lab partners/organization:	
Client:	Project:
Boring no.:	Recovery depth:
Recovery date:	Recovery method:
Soil description:	

II. TEST DETAILS

Initial specimen diameter, D_o:	Initial specimen area, A_o:
Initial specimen length, L_o:	Initial specimen volume, V_o:
Moist mass of specimen, M:	Dry mass of specimen, M_s:
Moisture content, w:	Total unit weight, γ.
Dry unit weight, γ_d:	Degree of saturation, S:
Membrane type:	Axial strain rate, $\Delta \varepsilon_l / \Delta t$:
Deformation indicator:	Force indicator:
Deformation conversion factor, K_L:	Proving ring conversion factor, K_P:
Cell pressure, σ_3:	Specimen preparation method:
Notes, observations, and deviations from ASTM D2850 test standard:	

III. MEASUREMENTS AND CALCULATIONS

Deformation Reading (G_L)	Axial Deformation (ΔL)	Load Reading (G_P)	Axial Load (P)	Axial Strain (ε_l)	Corrected Area (A)	Deviator Stress ($\Delta \sigma$)

EQUATIONS:

$$\varepsilon_l = \Delta L / L_o$$

$$A = A_o / (1 - \varepsilon_l)$$

$$\Delta \sigma = P / A$$

$$\Delta L = G_L K_L$$

$$P = G_P K_P$$

$$\sigma_{1f} = \sigma_3 + \Delta \sigma_f$$

σ_3:
$\Delta \sigma_f$:
σ_{1f}:

UNCONSOLIDATED-UNDRAINED TRIAXIAL TEST (ASTM D2850)
LABORATORY DATA SHEET

I. GENERAL INFORMATION

Tested by:	Date tested:
Lab partners/organization:	
Client:	Project:
Boring no.:	Recovery depth:
Recovery date:	Recovery method:
Soil description:	

II. TEST DETAILS

Initial specimen diameter, D_o:	Initial specimen area, A_o:
Initial specimen length, L_o:	Initial specimen volume, V_o:
Moist mass of specimen, M:	Dry mass of specimen, M_s:
Moisture content, w:	Total unit weight, γ:
Dry unit weight, γ_d:	Degree of saturation, S:
Membrane type:	Axial strain rate, $\Delta\varepsilon_l/\Delta t$:
Deformation indicator:	Force indicator:
Deformation conversion factor, K_L:	Proving ring conversion factor, K_P:
Cell pressure, σ_3:	Specimen preparation method:
Notes, observations, and deviations from ASTM D2850 test standard:	

III. MEASUREMENTS AND CALCULATIONS

Deformation Reading (G_L)	Axial Deformation (ΔL)	Load Reading (G_P)	Axial Load (P)	Axial Strain (ε_l)	Corrected Area (A)	Deviator Stress ($\Delta\sigma$)

EQUATIONS:

$$\varepsilon_l = \Delta L/L_o$$

$$A = A_o/(1-\varepsilon_l)$$

$$\Delta\sigma = P/A$$

$$\Delta L = G_L K_L$$

$$P = G_P K_P$$

$$\sigma_{lf} = \sigma_3 + \Delta\sigma_f$$

σ_3:	
$\Delta\sigma_f$:	
σ_{lf}:	